Coastal Tourism in Southern Europe in the XXth century

Patrizia Battilani & Carlos Larrinaga (eds.)

Coastal Tourism in Southern Europe in the XXth century

New economy and material culture

PETER LANG

Bibliographic Information published by the
Deutsche Nationalbibliothek
The Deutsche Nationalbibliothek lists this publication in the Deutsche
Nationalbibliografie; detailed bibliographic data is available online at
http://dnb.d-nb.de.

Library of Congress Cataloging-in-Publication Data
A CIP catalog record for this book has been applied for at the
Library of Congress.

Cover illustration:
Centro de Tecnología de la Imagen (CTI), University of Malaga. Archivo Arenas.
Reference: CTI AF0807_19630800_AR_2221_2613L701-I hotel triton 1963

Research project HAR2017-82679-C2-1-P, financed by the Ministry of Science and
Innovation of the Government of Spain and the ERDF.

ISBN 978-3-631-86471-5 (Print)
E-ISBN 978-3-631-86472-2 (E-PDF)
E-ISBN 978-3-631-86473-9 (EPUB)
10.3726/b18864

© Peter Lang GmbH
Internationaler Verlag der Wissenschaften
Berlin 2021
All rights reserved.

Peter Lang – Berlin · Bern · Bruxelles · New York · Oxford · Warszawa · Wien

All parts of this publication are protected by copyright. Any
utilisation outside the strict limits of the copyright law, without
the permission of the publisher, is forbidden and liable to
prosecution. This applies in particular to reproductions,
translations, microfilming, and storage and processing in
electronic retrieval systems.

This publication has been peer reviewed.

www.peterlang.com

Table of Contents

About the Authors 7

Introduction 11

Annunziata Berrino
Sustainability in tourism in Italy in post-World War II 19

Patrizia Battilani and Donatella Strangio
Mass tourism and social sustainability: Insights from the Italian and French coasts 37

Bertram M. Gordon
"Sous les pavés, la plage": Sun, sand, and surf in French tourism – The evolution of an image 57

Julie Manfredini
Transformations of tourism. On the French Riviera since the 1950s 77

Carlos Larrinaga
Spain after the Civil War (1936–1939). The new possibilities for maritime and coastal tourism 95

Marta Luque and Víctor M. Heredia
Pioneering projects in the tourism development of the Costa del Sol (Spain) 115

Margarita Dritsas
"White Flowers" of the Aegean. Would Le Corbusier use the same expression today? 133

Petra Kavrečič and Metod Šuligoj
"Socialist-style tourist accommodation" 155

Tomi Brezovec and Aleksandra Brezovec
Yugoslavia awaits you: post WW2 tourism promotion of the Yugoslav coast 171

About the Authors

Battilani, Patrizia. Full Professor of Economic History and, from 2018 to 2021, Head of the Center for Advanced Studies on Tourism of the University of Bologna (Italy). She has been visiting scholar at Sidney university and Glasgow university. Her primary research interests are the history of culture and tourism with applications in the field of participatory tourism planning, enhancement of UNESCO world heritage sites and European Cultural Routes, as well as business history with a focus on social enterprises. Currently she has the scientific responsibility for the Bologna unit of the Interreg Italy-Croatia Recolor (Reviving and EnhanCing artwOrks and Landscapes Of the adRiatic) and she is part of the research team of RurAllure (Horizon 2020), FabRoutes (Erasmus+), Mistral (Interreg Med). She is a researcher of the project "The tourism during the Civil War and the first Francoism, 1936–1959. State and private companies in the tourist recovery of Spain. A comparative perspective," funded by the Ministry of Science and Innovation of the Government of Spain and European Regional Development Fund.

Berrino, Annunziata. Professor in Contemporary History at the University Federico II – Department of Humanities in Naples and director of Centro interdipartimentale di ricerca sull'iconografia della città europea (CIRICE) of the same University. Her primary, though not exclusive, field of research is the history of tourism in the euro-Mediterranean area. She directs «Storia del turismo. Annale» (Franco Angeli editions) and «Eikonocity» (FedoaPress) and sits on the Advisory Board of the «Journal of Tourism History» (Taylor and Francis publisher). Among her books: *Storia del turismo in Italia* (Il Mulino, 2011); *I trulli di Alberobello. Un secolo di tutela e di turismo* (Il Mulino, 2012); *Andare per terme* (Il Mulino, 2014). She is a researcher of the project "The tourism during the Civil War and the first Francoism, 1936–1959. State and private companies in the tourist recovery of Spain. A comparative perspective," funded by the Ministry of Science and Innovation of the Government of Spain and European Regional Development Fund.

Brezovec, Aleksandra. Associated Professor of marketing and communication studies at the Faculty of Tourism Studies, University of Primorska (Portorož, Slovenia). Her research field connects tourism, communication studies and marketing. She has worked on several international research projects in the field of cultural tourism, tourism development and marketing. She is a member of

UNWTO Panel of Tourism Experts, a member of ECREA, and a member of UNESCO-UNITWIN Network / Chair "Culture, Tourism, Development" (Université Paris 1, Panthéon Sorbonne, France).

Brezovec, Tomi. Senior Lecturer at the Faculty of Tourism Studies, University of Primorska (Portorož, Slovenia). His main research interest focuses on tourism destination development. He is currently researching historical development of the Adriatic coastal resorts during the 19th and 20th century. The results of his research were published in several articles and book chapters on tourism history. He has been involved in several national and international tourism development projects and has curated several museum exhibitions on tourism development in Slovenia and Croatia. He is also a member of the advisory board of the Journal of Tourism History (Taylor & Francis).

Dritsas, Margarita. Professor of European Economic and Social History, Hellenic Open University, Patras, Greece. Has taught and directed seminars in Greek and foreign Universities on Greek and European Economic & Social History and History of Tourism. Member of the European Business History Association (EBHA); Member of the Advisory board of *Journal of Tourism History* (2009–2016); Honorary Member of the Hellenic Association of Women in Tourism. Publications on banking, business and tourism history. Recent works on Tourism: "Outline of Tourism in Greece during the Twentieth Century. Continuity and Change," *RHEE* 10–2016, pp. 53–81); Margarita Dritsas and Katerina Papadoulaki, *Ψηφίδες ιστορίας του ελληνικού τουρισμού (Selected Tiles from the Greek Tourism mosaic)*, Economia Publ. Athens (2019) (in Greek). Margarita Dritsas & Harry Coccossis (eds), *Tourism and Crisis in Europe XIX-XXI centuries. Historical, National, Business History Perspectives*. Economia Publishing, Athens (2014); 'Tourism and Business during the Twentieth Century in Greece: Continuity and Change' in L. Segreto, C. Manera & M. Pohl (eds), *Europe at the Seaside. The economic History of Mass Tourism in the Mediterranean*. Berghan Books NY, Oxford 2009.

Gordon, Bertram M. Professor Emeritus of History at Mills College in Oakland, California, specializes in modern French history and the history of tourism. His most recent book, War Tourism: Second World War France from Defeat and Occupation to the Creation of Heritage (Cornell University Press, 2018), explores the linkages between war and tourism. "Touring the Field: The Infrastructure of Tourism History Scholarship," in the Journal of Tourism History (September 2015), surveys the development of academic tourism studies, reflecting the growth of the industry, by some measures the largest single economic sector in the world. He is the author of Collaborationism in France during the Second World War (1980), based in part on interviews with former French supporters of Nazi Germany, and editor of The Historical Dictionary of World War II France: The Occupation, Vichy and the Resistance, 1938–1946 (1998). He has written on the 1968 revolts in France, the history of Vichy as a spa town, and is Associate Editor of the Journal of Tourism History. A co-editor of the digital discussion list H-Travel since its inception in 2003, he is also a core member of the Tourism Studies Working Group at the University of California, Berkeley.

Heredia, Víctor. PhD and Adjunct Professor of Economic History at the University of Malaga (Andalusia, Spain). His research fields are historical statistics, water supply history, history of education and industrial development in Andalusia in the 19th and 20th centuries.

Kavrečič, Petra. PhD, Assistant Professor at the Department of History at the University of Primorska, Faculty of Humanities, Slovenia. Her study field is economic and social history of tourism. She has published a scientific monograph and several scientific articles at home and abroad, co-edited monographs, focusing on the development of modern tourism in today's western Slovenia and the relation between tourism and political ideology on the case of commemorative practices of WWI battlefields. She has been and is involved in several research projects financed by the Slovene Research Agency. She conducts her research in domestic and foreign archives.

Larrinaga, Carlos. Associate Professor of Economic History at the University of Granada (Andalusia, Spain). His research is in the history of tourism, railways in the 19th century and the service sector. He is currently leading an interdisciplinary project on the history of tourism in Spain and Italy in the 20th century, funded by the Spanish Ministry of Science, Innovation and Universities and the European FEDER funds. He has undertaken research in several stays at Bordeaux-Montaigne University and at Aberystwyth University. He is the Main Researcher of the project "The tourism during the Civil War and the first Francoism, 1936–1959. State and private companies in the tourist recovery of Spain. A comparative perspective," funded by the Ministry of Science and Innovation of the Government of Spain and European Regional Development Fund.

Luque, Marta. Associate Professor of Economic History at the University of Málaga (Andalusia, Spain). Her research is in the history of tourism and in the labour market with gender perspective. She is a researcher of the project "The tourism during the Civil War and the first Francoism, 1936–1959. State and private companies in the tourist recovery of Spain. A comparative perspective," funded by the Ministry of Science and Innovation of the Government of Spain and European Regional Development Fund.

Manfredini, Julie. Doctor in contempory history, I specialized in history of tourism. Author of thesis on the history of *syndicats d'initiative*, published at PUFR, I discussed the place of *Côte d'Azur* in French tourism during my work. Associate researcher at EIREST (Paris I) and temporary worker at Paris III Sorbonne-Nouvelle. She teaches the history of tourism there.

Strangio, Donatella. Full Professor of Economic History at the Sapienza University in Rome (Italy). She is Director of the Cultural Heirtage Enhancement International Training Course at the Sapienza University. She is the scientific director of the Migrations/Migrations Series, Ed. Nuova Cultura, Rome. She is a researcher of the project "The tourism during the Civil War and the first Francoism, 1936–1959. State and private companies in the tourist recovery of Spain. A comparative perspective," funded by the Ministry of Science and Innovation of the Government of Spain and European Regional Development Fund.

Šuligoj, Metod. Associate Professor at the Department for Management in Tourism at the Faculty of tourism studies – Turistica, University of Primorska, Slovenia. His area of research relates to dark tourism, special interest tourism and hospitality management. He often blends and bridges theory from a variety of disciplines (e.g. management, different types of sociology, geography, history) in order to explain social phenomena in tourism, especially in the Balkan and Upper Adriatic context. He utilizes qualitative and quantitative methodology in delving into the complex social phenomena associated with tourism.

Introduction

Since the pioneering studies of J. Walton (Walton, 1983) in the UK, A. Corbin (Corbin, 1988) in France and G. Triani (Triani, 1988) in Italy, the history of tourism has provided knowledge, interpretations and also material for new museums and collections. Through historical analysis, the civilization of seaside tourism has found its own narrative and recognition. In this context, the Mediterranean countries have been intensively investigated due to their prominent role in European tourism. There is a long list of good essays and articles focusing on the driving factors as well as the economic impact of tourism in individual countries and destinations. However, only a few books and special issues of journals have tried to provide an overview or a comparative analysis of the Mediterranean countries. Of these few we can mention *Europe at the seaside* by L. Segreto, C. Manera & M. Pohl (Eds.) in 2009 (Segreto, Manera, Pohl, 2009) and issue n. 10 of the *Revista de la historia de la economía y de la empresa* in 2016, entitled "El Mediterráneo: mucho más que sol y playa (1900–2010)." (Hernández Andreu, 2016) Finally, last year (2020) the Journal *Transportes, Servicios y Telecomunicaciones* (Larrinaga Rodriguez, 2020) dedicated the entire issue n.41 to a comparison of Italian and Spanish tourism from World War I to the 1950s.

This book is motivated by the desire to provide an overview of the development of tourism along the Mediterranean coasts on the basis of the new cultural approaches that have emerged over the last decades and the new challenges facing our civilization. The "sustainability turn" and the "material turn" are the lenses through which we can read these essays, which analyse how maritime and coastal tourism gave birth to a new economy and material culture in the20th century on the Mediterranean coasts, focussing on different countries and regions in southern Europe: Spain, France, Italy, former Yugoslavia, and Greece.

When the first books on the history of tourism were published in the 1970s and 1980s, a negative view of this social and economic phenomenon prevailed among scholars. The impact of interpretations by prominent sociologists, philosophers and town planners was fundamental, such that tourism was often considered a by-product of the consumption society, devastating beautiful landscapes, the natural environment and cultural heritage in order to provide amusement for uncultured people (Enzensberger, 1958; Fink, 1970; Poon, 1993; Urry, 1990). The first generation of tourism historians we mentioned at the beginning of this section paved the way for a profound change and the recognition of tourism as an important economic industry contributing to the improvement of the quality of life of the middle and working class. In reality, in the 1960s some international organizations also began to look differently at these economic activities. Take for instance the United Nations conference held in Rome in 1963, which "laid particular stress on the social and cultural value of tourism" and "recognized that tourism was important not only as a factor of foreign exchange but also as a factor in the location of industry and in the development of areas poor in natural resources" (United Nation, 1963, p.15). At the same time UNESCO began to include tourism in the economic activities that were able to support the preservation and enhancement of cultural heritage. Furthermore, in 1969, the World Bank set up the Tourism Project Department which three years later launched a tourism sector strategy aimed at funding tourism development plans in developing countries. This department was then closed in 1979. Finally, the World Tourism Organization, initially a network of entities operating in tourism promotion and development, started to move toward a wider vision by designing the role of tourism in the transformation of our society. This was the beginning of a more grassroots process that facilitated the approval of international conventions and charters promoting sustainable tourism and assigning tourism an important role in social, cultural and economic development. The three pillars of sustainability (economic, environmental and social/cultural) are actually the most important challenges for tourism planners and policy makers and should also intersect with the historical aspects.

Sustainability can be considered the first important "turn" in the history of tourism. For this reason, 3 essays in this book deal with the involvement (or empowerment) of residents in tourism planning (social sustainability) and beach and water management (environmental sustainability).

The essay *Mass tourism and social sustainability: some insights from the Italian and French coasts* by Patrizia Battilani & Donatella Strangio analyses social sustainability in the context of tourism history by focussing on the ability of DMOs to involve local stakeholders in designing tourist destinations in three Italian and French regions. It is well known that the three main components of social sustainability are social justice, participation in the decision-making process and preservation of minority cultures. The DMOs strongly contributed to making mass tourism socially sustainable, as they were a tool for meeting and representing the various stakeholders. In some territories they also stimulated consideration of crucial problems for the growth of tourism: in the cases of Rimini and Riccione they committed themselves to developing the airport and charter flights (in collaboration with the EPT), and led many initiatives aimed at raising stakeholder awareness of the need to build positive relationships with tour operators and upgrading hotels to the standards they required. By focussing on stakeholder participation in tourism planning, the authors reassess the sustainability of mass tourism.

The essay by Annunziata Berrino, *Visions, projects and sustainability in tourism in Italy in post-World War II*, draws attention to environmental sustainability, analysing a set of local disputes that arose over environmental resources in the late 1940s and 1950s in Italy and accompanied the spontaneous phase of mass tourism. At the time the role of local intermediary institutions was very important, as they were able to communicate with the central government bodies in Rome – which were undergoing a comprehensive process of reorganization after Fascism – and to interpret local dynamics in the delicate reconstruction and recovery phase. In this context the tourism boards (nowadays we would call them destination management organizations (DMOs)) were definitely committed to protecting the cultural and environmental resources of their areas of responsibility. Throughout the 1950s, while coastlines were quickly being transformed to meet the needs of seaside tourism, the conflict between the private sector and peripheral tourism bodies was also reflected in disagreements centred around the use and safeguarding of natural and environmental resources that were useful for tourism.

Julie Manfredini's essay *Transformations of tourism. On the French Riviera since the 1950s* focusses on the French Riviera, where tourism accentuated difficulties like water shortages in summer and coastalization and contributed to serious pollution problems such as erosion. However, the experience of local tourism actors has enabled the establishment of policies for the regulation, management and preservation of the environment, especially since the 1960s, when several coastline preservation projects were developed, such as the Vaugrenier department park in the Alpes-Maritimes. Some cities have developed their pedestrian areas in city-centres since 1977 and others, for example Nice, refused so-called "savage urbanisation" and retained the "longue tradition de régulation urbanistique." Extensive artificialization of the coast has been observed, largely due to the maritime attraction exerted by the coast and heliotropism. The construction of housing and port infrastructure stems from the desire for views of the sea and this infrastructure has saturated the coastline.

The third pillar of sustainability is linked to the economic model. Economic sustainability refers to viable, long-term economic activities providing fairly distributed socio-economic benefits to all stakeholders, including stable employment and income-earning opportunities and social services to host communities, and contributing to poverty alleviation (UNWTO, 2019). Four essays explore the impact of tourism on the local economic fabric. Víctor Heredia & Marta Luque and Carlos Larrinaga in two separate essays on Spain, and Margarita Dritsas on coastal tourism in Greece all analyse the role played by local entrepreneurs and institutions in designing a viable and long lasting tourism sector, including the role of international capital and recovery plans.

Víctor Heredia & Marta Luque in *Pioneering projects in the tourism development of the Costa del Sol* (Spain) focus on the the first half of the 20th century in analysing how the Costa del Sol gradually established itself as a tourist destination. The paper highlights the role played by the foreigners' colony which established itself there at the turn of the century and the role foreigners played in supporting the most important initiatives of the time such as the first hotel, new attractions (golf) and transport infrastructures (the airport), which are still important drivers of tourism on the Costa del Sol. However, the essay also describes a proactive local community which rapidly took the lead in tourism planning, although some of the local private entrepreneurs' projects, which were equal in complexity to those of the main European tourist leisure centres, were frustrated by different circumstances.

The essay by Carlos Larrinaga, *Spain after the Civil War (1936–1939). The New Possibilities for Maritime and Coastal Tourism* analyses the origins of coastal tourism in Spain from the second half of the 19th century onwards. The first contact with the coast, from the tourism point of view, was related to hygiene and health. It was only at the end of that century when medical science created more effective drugs than water intake, that the healthy character was replaced by an increasing sense of enjoyment. Hence, holidaying at the seaside gained importance among the most privileged groups of society. Yet it was only after World War II that the definitive transfer of tourism flows to the Mediterranean occurred, becoming the favoured destination of hundreds of thousands of holiday makers. Spain, which until then had an insignificant role in international tourism and received a rather small number of foreign tourists compared to France or especially Italy, began to see its beaches on the island of Mallorca and on its Mediterranean coast filled with tourists. The development of car and air travel had a great impact on these movements. Also this essay highlights the intersection between the local community and foreigners in designing and promoting the "Spanish tourist miracle"

The chapter by Margarita Dritsas entitled *"White Flowers" of the Aegean. Would Le Corbusier use the same exclamation today?* analyses the development of coastal tourism in Greece and challenges the traditional interpretation that early tourism development occurred in the 1960s. The essay shows that tourism planning dates back to the end of the World War II, when a joint rescue plan for Greece was negotiated by US and Greek authorities within the framework of the Marshall Plan. For the first time tourism was recognized as the third most important autonomous economic sector. Implementation of the plan depended on the Greek bureaucracy, special institutions, and the rather belated spontaneous and unregulated response of wider society which shared a particular perception of foreign tourism. From the beginning, therefore, sea and coastal tourism was seen as a low-cost development/investment process which would attract large numbers of tourists and secure high enough state revenues. Despite the many lacunae in the conception of the programme and the ignorance of Greece's culture and structure by foreign advisors, as well as the inadequacies of Greek state bureaucracy, the deficiency of political personnel at the time and world political priorities set by the Truman Doctrine related to the Cold War, the programme did contribute to restoring the Greek economy by promoting tourism, especially after the monetary stabilization of 1954 (devaluation of the drachma) which made Greece a very cheap country to visit.

The second "turn" we address in the book is the "material turn," that is, historians' engagement with material culture and its methodologies in asking new questions, dealing with new themes, analysing new sources and stimulating collaboration between historians and practitioners (Gerritsen & Riello, 2014).

It is well known that for much of the 19th century and even into the early 20th century, the prevailing hygienist paradigm made the European Atlantic façade the favourite tourist destination of the upper classes of European society. At the same time some Mediterranean coastlines began to cater to local and national tourists. However, since the beginning of the 20th century the process of "going south" has become progressively stronger, making the warm beaches of southern Europe a sort of capital for international tourism. The Mediterranean and tanning became fashionable, while sports such as swimming were more suitable for the warm waters of the south of the continent. The arrival of tourists changed the use of beaches as well as the traditional view of the sea. New associations were set up to promote new lifestyles and sports activities such as for instance yacht clubs. People taking a walk, practicing sports or sunbathing progressively populated desert beaches. New professions as well as new enterprises emerged to provide a variety of services aimed at welcoming and entertaining people on the beach or in the sea: lifeguards, beach service attendants, skippers, etc. New goods started to be produced and consumed such as beach umbrellas, beach huts or cabins, deckchairs, etc.

The passage from a hygienist view of the beach to the sun, sand, surf model (or spirits and sex, depending on the place) is investigated for France by Bertram

Gordon, who in his essay, *"Sous les pavés, la plage" – Beach and Seaside Tourism in France since the Second World War*, sums up the history of sun, sand, and surf tourism along the beaches of France in the 20th and 21st centuries. Films such as Maurice Dekobra's "Le Train Bleu" and novels, notably *Tender Is the Night* by F. Scott Fitzgerald, burnished the tourism imaginaries of the Riviera during the interwar years. Of equal, if not greater, importance were shifting attitudes toward skin colour, with pale replaced by a growing fashion of *bronzage,* or sun-tanning, among the upper and middle classes, beginning in the early 20th century. This shift, related to changing perspectives about the medicinal benefits of the sun, as well as evolving cosmetic fashions, both discussed in this essay, is a striking example of the intersection of tourism with other cultural trends in history.

A different aspect of the material culture is investigated by Petra Kavrečič & Metod Šuligoj in their essay, *"The socialist-style tourist accommodation"* in the eastern part of the Upper Adriatic. The aim of the paper is to illuminate the specific Yugoslav model of domestic tourism and related "socialist-style tourist accommodation," which were primarily intended for the working class, a pillar of the then socialist society. After the Second World War the new Yugoslav administration transformed the organization of tourism, which until then had been mainly based on private activity, into a predominant social sector. Thus, large state-owned tourism companies first managed the nationalized tourism facilities and later built new ones. The main investors ("owners") in workers' tourism were Slovenian continental state-owned companies and the ministries for public sector employees. Slovenian investors invested not only in the Slovenian coast, but also elsewhere in the eastern part of the Upper Adriatic, including the rest of the Istrian peninsula.

This transformation also had a "material cultural" dimension with the diffusion of a special type of accommodation opened all along the Slovenian coast in the form of holiday homes of various state-owned companies. The "socialist-style tourist accommodation" was a very non-homogeneous coinage that included very different types of accommodation establishments (i.e. private holiday homes and those for workers, hotels, apartments, bungalows, and campsites, individual pitches in campsites), which were also used differently by individual groups within socialist society. However, the Yugoslav socialist model based on the working class was obviously not entirely consistent as elites nevertheless holidayed (rested and relaxed) differently than the working class. In practice, this meant the coexistence of two models, one based on the left social ideas and the other which was clearly more similar to those of western capitalists; these models did not include the same type of accommodation. The Yugoslav socialist myth of equality and solidarity was certainly violated in the sphere of holiday practices.

The essay by Tomi Brezovec and Aleksandra Brezovec, *Yugoslavia awaits you: post WWII tourism promotion of the Yugoslav coast*, takes into consideration another aspect of the material culture which is the representation of objects in advertising and promotional campaigns. In the course of about two decades of post-war tourism development, Yugoslavia developed into a meaningful player on the tourism market. During the 1950s, after Yugoslavia had overcome its early difficulties in the process of establishing of a new, rather unique political and economic system, it started to accept tourism as a viable economic activity. At first the endeavour in tourism was cautious if not suspicious, and was not without political connotations. Tourism brochures for foreign visitors promoted the county's nature and history but also its political and economic system. They mentioned the fight for freedom and the prized achievements of the socialist society. Gradually, as the country assured its position on the political map, the need for an ideological context in tourism promotion diminished. Tourism promotion shifted from serving the political agenda to serving the commercial interests of the industry. In the 1960s, in the process of the massification of tourism, Yugoslavia positioned itself as a cheap country for seaside summer holidays. Brochures, now cheaper and smaller in size, were limited to selling "sea, sand and sun."

In conclusion, the book stimulates thought on a new narrative of tourism history that is able to face up to society's new challenges, namely sustainability in all its dimensions and an understanding of the role played by tourism experiences in the Western material culture.

Patrizia Battilani & Carlos Larrinaga

References

Corbin, A. (1988), *L'Occident et le désir de rivage 1750–1840*. Paris: Aubier.

Enzensberger, H.M. (1958), Vergebliche Brandung der Feme: Eine Theorie des Tourismus. *Merkur* (126), https://www.merkur-zeitschrift.de/hans-magnus-enzensberger-vergebliche-brandung-der-ferne/

Fink, C. (1970), *Der Massentourismus*, Bern: Verlag Paul Hupt.

Gerritsen, A. & Riello, G. (2014), *Writing material culture history*. London: Bloomsbury Academic.

Hernández Andreu, J. (2016),"El Mediterráneo: mucho más que sol y playa (1900-2010),"*Revista de la historia de la economía y de la empresa*, issue 10.

Larrinaga Rodriguez, C. (2020), "El turismo en la Edad contemporánea.Un fenómeno transnacional", *Transportes, Servicios y Telecomunicaciones*, issue 41

Poon, A. (1993), *Tourism, Technology e Competitive Strategies*, Wallingford: Cab International.

Segreto, L. Manera C. & Pohl M. (Eds.) (2009), *Europe at the seaside: the economic history of mass tourism in the Mediterranean*, New York, NY: Berghahn Book.

Triani, G. (1988), *Pelle di luna pelle di sole. Nascita e storia della civiltà balneare 1700-1946*. Venezia: Marsilio.

United Nation. (1963), Recommendations on International Travel and Tourism, Conference held in Rome (31 August to 5 September 1963)

Urry, J. (1990), *The tourist gaze*, London: Sage

Walton, J. (1983) *The English seaside resort: a social history, 1750-1914*. Leicester: University Press.

Annunziata Berrino

Sustainability in tourism in Italy in post-World War II

Abstract In post-World War II in Italy the theme of tourism returned to the public debate, after the twenty years of fascism. On the one hand, the exponents of industrial capitalism expressed their vision of the future of Italian tourism and will propose important investments to support their ideas, also in relation to industrial development. On the other hand, the small family business intensified spontaneous and often disordered development processes. The chapter analyses the debate on environmental sustainability that accompanied the two models in Italy in the years between 1945 and 1970.

1 Introduction

In the context of the history of tourism, authors acknowledge that active tourism associations, such as the Touring Club or the Alpine Club, have played a major role in Italy, as in the rest of Europe, in developing and extending a culture of safeguarding cultural and environmental heritage ever since the early 20th century (Piccioni, 1994; Moranda, 2015) but that this constructive and proactive commitment has only minimally penetrated into the common sentiment, so that Italians may be defined as "destroyers" of the landscape and the environment (Pivato, 2006, p. 153). Environmental historians, however, unanimously support the theory that Italy has never given due concern to protecting its environmental heritage, in particular in the tourism massification phase, that is, in the second half of the 20th century (Corona, 2015), pursuing "a territorial administration plan that is increasingly indifferent to landscape considerations" (Lanzani, Bolocan Goldstein & Zanfi, 2015, p. 291), and to the environment in general; on the contrary, "in Italy's material constitution, the landscape has not been acknowledged as a common good over the years, but as a positional good to be "jealously" guarded by individuals or "democratically" exploited to the point of exhaustion and decay" (Lanzani, Bolocan Goldstein & Zanfi, 2015, p. 295).

This essay offers an analysis of certain moments in the history of tourism in Italy during the years from the end of World War II to the early post-war years and, more precisely, from 1945 to the end of the 1950s, when a series of transformations reversed the opinion of tourism from being the right way to safeguard the cultural heritage to an activity that had to be regulated, due to its detrimental impact on lands and communities (Zuelow, 2016).

That fifteen-year period is generally neglected: research has focused more on analysing the 1960s that followed, when Italy was remarkably successful in the international tourist markets and when, as a consequence, there was a real assault on the coasts and rampant real-estate speculation in coastal resorts. In fact, it was in the early 1960s that the opinion of tourism changed: it was no longer a development opportunity for which it was important to safeguard natural and cultural resources, but was added to all those activities that threatened the environment. Thus an explicit culture of condemnation also started to materialize in the context of tourism, blaming both the State and private individuals in equal measure (Pivato, 2006, p. 149).

Analysing the years preceding the tourism boom is nevertheless important for understanding the political, institutional and cultural conditions that prevailed in Italy, which was soon to suffer from the rapaciousness of the massified dimension of tourism. These were crucial and difficult years, during which Italy was working hard to rethink its cultural and environmental resources to adapt them to the common images and tourist practices of the post-war years, as the country was becoming accepted into the western international political landscape (Berrino, 2011). At the same time, it was having to deal with huge battles internally in redefining its own governance and tourist policies, in order to overcome the elaborate ones imposed by the fascist dictatorship.

2 Between safeguarding "in favour of" tourism and safeguarding modern life

In 19th-century Italy, tourism was initially only rarely opposed or criticized as it developed and expanded. The economic advantages that local communities were gaining from the movement of visitors in transit, passing through or staying for a holiday substantially compensated for the effects of any inconvenience. Greater mobility of people somewhat increased phenomena such as begging, prostitution or small-scale delinquency, which were almost a commonplace in the reports and accounts of travellers in the more popular town centres, or in the more isolated regions of the south where tourism had also encountered social unease, perhaps manifested simply as banditry (Fiore, 2018).

However, as far as the use of resources is concerned, countless studies show that in many places one of the most frequent reasons for complaint, expressed by visitors themselves, by operators or local communities, was the forced collective use of common resources by different populations, both healthy and sick.

These contrasts were nevertheless resolved by creating, for example, physically separate facilities, visibly recognizable for healthy tourists and for sick tourists. The phenomenon appeared almost everywhere, both in seaside bathing places (Kavrečič, 2016), and in high mountain resorts, or spa resorts (Large, 2019). In any case a balance was reached, as the explicitly therapeutic nature of the earliest tourist practices had anyway led to the creation of environmental and urban renewal interventions, designed according to theories of hygiene, which were of undoubted advantage for the areas. The fact that the exploitation of natural resources in water and bathing resorts during the 19th century produced administrative disputes that were often interconnected with very long, intricate political controversies is well known, but at this stage, it appears that the questions were posed and resolved according to a logic dominated by profit.

Towards the end of the 19th century, the early expansion phase of globalization had encouraged an awareness of both the value and the effects of tourism, which was starting to be related and compared to industry and modern life in general. These were extremely interesting dynamics, often simplified, but which the recent history of tourism linked with the history of the environment are revealing in their complexity (Moranda, 2015).

In this period, European associations, including the Italian Touring Club, posed the question of protecting natural resources, landscapes and cultural heritage, and the arguments were generally in favour of tourism and for encouraging tourism, as the coasts, mountains and waters and the vast cultural heritage were considered the heart of the nation's tourist attractions and identity. So at the turn of the 20th century, the need to pay attention to the Italian landscape and to historical contexts was expressed more and more clearly and led to the first standards for safeguarding, such as the uniform organic Law no. 364 of 1909 on safeguarding the cultural heritage, "For antiquities and the fine arts," also known as the Rosadi-Rava law. In the years between the two world wars, during the fascist dictatorship, there was a dominant tendency to modernize environments, the most evident example of which was the capital, Rome (Gentile, 2007). Nevertheless, in spite of censure, the development of a critical attitude towards more aggressive modernism, and the reinforcement and extension of the concept of safeguarding continued, leading to Law no. 778 of 11 June 1922, "Measures for the protection of natural beauties and buildings of particular historical interest," and almost taking on features of an ideology at times. An example of this is the case of the urban "trulli" settlement at Alberobello, whose integral preservation was even reinforced by a special decree of 1931 that, while preserving for future generations a priceless cultural heritage, recognized by UNESCO in 1996, in actual fact it restricted the vital needs of the resident population (Berrino, 2012).

Finally, Laws no. 1089, "For the protection of objects of artistic or historical value" and no. 1497 "For the protection of natural beauties" were passed in 1939.

Until the mid-20th century, the defence of cultural and environmental heritage by the modern world generally – and therefore also by tourism – was expressed on several sides, and individual sensitivities, the speeches and the arguments were often at odds with political thinking and cultural affinities.

3 Protecting environmental resources in the territories at war

As briefly mentioned above, at the outbreak of the Second World War Italy already had some experience of managing pressures on environmental resources and social conflicts generated by tourism on communities, and these experiences were useful in tackling the most critical moments of the war.

In Italy, the most serious devastation occurred in 1943: the Italian government had, in fact, signed the armistice and the German armed forces, no longer allies but enemies, retreated from Italy pursued by Anglo-American forces who had disembarked on the southern coasts. The bombing damaged or destroyed basic tourist facilities, such as hotels and bathing resorts, but in some places military occupation also threatened the integrity of certain natural environments. In any case, the available documentation shows that communities were aware of their own resources, which they considered to be strategically important for the post-war economic recovery. It is worth reconstructing some of these events.

The liberation of Italy from Nazism started in the regions of the South. In Campania, at Amalfi – already liberated by American troops in the autumn of 1943 after the fall of fascism – the local Tourist Board was managed by Ruggiero Francese, an engineer and a communist, who had a very clear vision of tourism and the potential of the entire Amalfi coast; the Amalfi coast was to be included in the list of UNESCO World Heritage Sites in 1997. The Amalfi urban development plan provided for interventions for hygiene, as well as freeing up the beach and the sea-front to meet the increase in demand for sea-bathing, and also moving the naval carpentry yard and the fishing gear storage depot to other areas. In addition, it was Francese who, in the 1930s, had fought for the promotional development of a cave on the coast, the Grotta dello Smeraldo, which was similar to the famous Grotta Azzurra on the nearby isle of Capri, accessible only by sea and with a wealth of stalagmites. The cave was opened to visits from the Anglo-American troops, in order to show "foreigners, also those in military uniforms" the wonders the Amalfi coast could offer to world tourism (Francese, 1949). Even more interesting is the fact that the local tourist office secured an agreement with Federterra (Federation of agricultural labourers) and together they founded a cooperative with the aim of assigning the coastal uplands to farmers, thus responding to the problem of agricultural unemployment. Francese even encouraged the opening of "elite" tourist villages on the hilltop area. It was a vision that protected the landscape and tried to safeguard the fairly fragile balance of the whole of the high, rocky coastline. For Francese everything had to be done "always (…) in defence of the landscape, our sacred heritage," even though an almost general insensitivity prevailed all around. "We must defend Amalfi, which has immense tourist possibilities. We must understand tourism as the most efficient means for an awareness, understanding and solidarity between peoples, and therefore for peace in the world" (Amalfi (Salerno). Archivio dell'Azienda autonoma. Amalfi, 8 settembre 1947. Relazione del Presidente uscente ingegnere Francesco Ruggiero al Consiglio direttivo dell'Azienda).

In the case of Amalfi, the upheavals of the war and the presence of foreign military forces were therefore an opportunity to reorganize and safeguard the coastline and to promote the discovery of natural surroundings, such as sea caves.

However, a lot of serious damage was caused by the double military occupation – first of all, German, and then Anglo-American – in the tourist resorts of Central Italy, and in particular the spa resorts which, as they were better equipped with hospitality facilities and other services, were used for logistics, as also happened in the rest of Europe (Large, 2019). The town of Montecatini in Tuscany, which had 100,000 visitors a year before the war, including many foreigners, was used by the Germans as a hospital centre. With the arrival of the Allies, the occupation was even more widespread (Ruata, 1946b), since they used it as a distribution hub, periodically stationing 20,000–30,000 military personnel there. As soon as they arrived, the allied army began building huts containing bathrooms and pit latrines, and these works gave rise to "deep concerns" over the serious danger of infiltrations and pollution they might cause to the water catchment area. Damage of this kind could have paralysed the work of the Montecatini spa businesses for years. Continual protests and warnings led the Rome government to intervene with the military high command to avert any damage. The proposal was then made to designate Montecatini as "a place of convalescence for allied forces' military personnel affected by gastro-hepatic and gastro-nervous ailments contracted in war"; the place was, in fact, designated as being for hospital and distribution use, and operated as such until early September 1945 (Roma. Archivio centrale dello Stato. Presidenza del Consiglio dei ministri. Atti di gabinetto. 1944/47–1944, fascicolo 1.6.1, 17898).

Further north in Veneto, Abano was also an important spa town, with 25 hotel establishments and a total of 1,325 rooms. These facilities could stay open the whole year round, with their heating systems that exploited the heat of the spa waters. Before the war, Abano had 40,000 visitors a year, including 5,000 from abroad. At the beginning of October 1943, it was evacuated and entirely requisitioned by the Germans, who installed the general headquarters of the Luftwaffe, the German air force, there. In April 1944, after the German withdrawal, Abano was also transformed into a hospital centre for British troops, and this remained active until mid-1946. Abano was saved from destruction, but its thermal mud deposits were used as rubbish dumps. Private firms in Abano, too, were ready to tackle the recovery of natural resources and the reconstruction "with dogged determination" and they were supported in this by the local tourist office (Ruata, 1946a, p. 21).

4 The questions of the early post-war period

As the above events show, the local tourism institutions had an important role to play in protecting not only real estate assets, but also natural resources in the more difficult stages of the war; on many occasions, they had an active role and fought alongside communities and operators to defend the heritage – both cultural and environmental – on which tourism was based. In Italy, in the years between the wars when more liberal ideas prevailed, a large peripheral institutional bureaucracy was formed. From 1926 onwards, the government accepted the creation of local tourist offices in internationally-renowned tourist resorts; these were mixed public and private bodies, liberal in their conception, based on the French model of the *Syndicat d'initiative* and the *Kurkommissionen* in Austria and Germany. In 1932, the dictatorship also installed a provincial Tourist Board in every province, to control and administer the movement and the tourist business throughout the entire territory of Italy. A bureaucratic structure was formed within these bodies, which was well versed in local dynamics and the technical questions affecting tourism.

In the debate concerning reconstruction that took place immediately after the war, the Italian Touring Club believed strongly in this peripheral bureaucracy; it considered that these institutions should actually create local study centres to monitor building renovation and safeguarding of the environment, by means of projects, exhibitions and photo catalogues In the bombed towns especially, reconstruction should not be limited to saving monuments and rebuilding hotels, but their environmental framework should also be preserved. This was the opinion of Giuseppe Silvestri (1899–1973), among others. A journalist from the Veneto and an antifascist, he was also a supporter of the Touring Club and an attentive observer and defender of the environment; he feared that, after being subjected to aerial bombings, towns might lose their characteristic aspects in the rush to rebuild them.

> Too many experiments have been carried out on the corpses of our old towns, senselessly and mercilessly tearing them apart, for us to restore the same systems, and repeat the same errors, after the recent destruction caused by the war (Silvestri, 1946, p. 26).

Silvestri hoped the new urban development plans and reconstruction projects would be fully shared and discussed before being approved, and the appropriate location for this could be the local tourist bodies.

Anyway, when fascist censorship was lifted and the war was over, and before the political parties defined and took up their positions on tourism, the world of tourism had a voice on the new international scene and brought into play all its skills at conferences and in publications, identifying the more serious critical points and, in many cases, indicating the solutions.

These were local government secretaries, directors or chairmen of local tourist offices and provincial Tourist Boards who had had practical experience of very complex matters and who were able to follow and see connections between new arguments in a theoretical debate, at the highest level, on protecting the Italian cultural and environmental heritage.

It seems inevitable, then, that in the first post-war public meetings, for example, the first national tourism conference organized in Genoa by the local Chamber of Commerce in 1947 (Camera di commercio, industria, artigianato di Genova, 1947), the representatives of the territorial tourist bodies proposed intervening urgently to regulate the concessions, uses, exploitation and investments on the coastline. In the years between the wars, in fact, tourism practices and economic interests had been quickly concentrated on the seashores, and those who operated them were well aware that post-war tourism would start from there, even though government bureaucracy was looking elsewhere, to the great cities of art and to the expected copious influx of Americans.

However, the sea coasts were to be the real centre of attention.

The different nature of the Italian coasts, the long history of tourism in certain resorts – whose development dates, in some cases, right back to the mid-17th century – the times and outcomes of investments, and the forms of use and promotion posed a wide variety of questions.

Generally, the local institutions, being mixed public-private in nature, were familiar with the dynamics and conflicts generated by tourist occupation of the coasts and beaches, and they asked above all for regulation and planning tools.

The Salerno Tourist Board explicitly contested the empiricist and spontaneous attitudes that had been adopted up to that time, and considered that the post-war reconstruction should be carried out by technicians. The local boroughs, in fact, did not have urban development plans that would take into account the aesthetic and hygiene requirements of the resorts and, for example, they had allowed the construction of hotels in proximity to factories or in unsuitable or rundown areas.

> We must be convinced of the need for large-scale industrialized tourism, resting on solid technical criteria that must eliminate all the existing gaps and eyesores to offer complete, perfect hospitality. It is therefore essential to proceed to tourist planning of all the major centres in Italy. This planning must be organized and looked after by the provincial Tourist Boards, and entrusted to technicians of undisputable expertise, who specialize in town planning (Ente provinciale per il turismo di Salerno, 1947, p. 72).

Many municipal administrations aspired to encourage tourism without having the essential services, such as roads, aqueducts, sewage systems or cemeteries. In these cases, the local provincial Tourist Boards, together with the mayors and public works offices had to act as a mouthpiece to ask the State to create the necessary works. The methodology was finally decided on: technicians were not expected to look after hospitality facilities, but only the general conditions of the area:

> our tourist towns and villages should be renewed, although maybe gradually. (…) we must make our towns and villages look respectable, eliminating all the eyesores, all the inconsistencies and all the incongruities they are full of (Ente provinciale per il turismo di Salerno, 1947, p. 74).

The provincial Tourist Boards, along with the local borough councils, had to make all kinds of financial efforts to create "planning" activities; the State was asked to make the drawing up of an urban development plan compulsory for all local boroughs of tourist interest (Ente provinciale per il turismo di Salerno, 1947, p. 74).

The request to the government, repeated on several sides, to adopt an overall view of tourist areas and centres was supplemented by specific questions posed by the seaside resorts.

First of all, the urgent need to protect the beaches. The appeal for this was made by Abele Ciccaglione and by Silvio Volta, respectively secretary and chairman of the Savona provincial Tourist Board in the Liguria region, on the northern Tyrrhenian coast.

The quarrying and consequent removal of sand, gravel and shore from the Italian coastline was governed by the Merchant Shipping Code: in some places, quarrying was free, in others a permit was needed from the maritime authorities, while in others again it was absolutely forbidden. The beaches were then classified differently and the classification was carried out by the Port Authorities and by the Public Works Offices, who decided where and how material useful for building could be obtained. In the post-war period, technicians asked that the classification of beaches as quarries should be done after consultation with the world of tourism, meaning after also hearing the opinion of provincial Tourist Boards, local tourist offices and the borough councils involved.

> This request for measures was justified by the ever-increasing and frightening erosion taking place on the beaches of many coastal areas of Italy and especially those of the Ligurian Riviera, subject to a high level of corrosion, and by the fact that the Public Works office thinks above all of the pressing and essential needs of construction. During this early post-war period, in fact, we have had to witness the "assault" – and it is, without exaggeration, an "assault" – on the beaches of our rivieras to grab sand, which is then accumulated and stockpiled by speculators as if it were grain (Ciccaglione, 1947, p. 179).

Rapid urbanization and construction systems required larger and larger quantities of sand and gravel, which were taken from beaches or from river beds near the estuaries even before the alluvial process could begin, and this produced the shortening or disappearance of beaches. To combat this phenomenon, cliffs were constructed, perpendicular or parallel to the coast, and so the beaches had three forces opposing them: the quarrying of sand and gravel for use in the building industry; the attempt to repair the damage by constructing cliffs; and the cliffs that were built by the railway companies to protect the coastlines.

While the demand for beaches and coasts was increasing, the bathing resorts, especially those near the large, fast-growing industrial urban centres, were trying to have their concerns heard by the State authorities. In Liguria, the Savona tourist board succeeded in reaching an agreement through the mediation of the Prefecture, but the local technicians were not seeking improvised solutions: they wanted to be considered as fixed interlocutors in the management of the coasts and asked for the old classification of the beaches as quarries to be reviewed in the light of the new interests of tourism (Ciccaglione, 1947).

So, the state administrations in charge of conservation of beachfront resorts were subject to the pressures of the building industry, which defended its work on behalf of the number of people it managed to employ; on the other hand, the technicians of the local tourist offices claimed "the duty of defending tourist interests, directly linked to the conservation of beaches," and they even went as far as comparing the tangible and intangible value produced by construction with that generated by tourism, and concluded that quarrying sand from coastal beaches brought about "a colossal destruction of wealth, disproportionate to the insignificant benefit that the building industry draws from it" (Volta, 1947, p. 211).

In those early post-war years, major backing for local dynamics in defence of natural and landscape resources for tourism also came from the Senate parliamentary group for tourism, which was very active in supporting policies for international cooperation. The group was coordinated by Luigi Gasparotto (1898–1954), a committed figure in numerous civil and political battles – defence of workers' rights, women's right to vote in local elections, the right to Sunday rest, and the protection of war veterans: he was an opponent of fascism and living in exile in Switzerland (D'Angelo, 1999). Gasparotto had already been active in defence of the countryside during the fascist period: in 1925 he asked the government to intervene urgently in accordance with the 1922 law to defend the "few garden areas" that still existed in the big cities, and in order that, as well as the supervision carried out by the regional supervisory bodies for monuments, local borough councils should be sensitized to intervene "promptly to save the threatened heritage," in particular, by purchasing the last free private areas to prevent them being the subject of urban speculation (Sesto San Giovanni. Fondazione Istituto per la storia dell'età contemporanea. Fondo Luigi Gasparotto. Unità 51 – Corrispondenza di Luigi Gasparotto e ritagli stampa sulla tutela del paesaggio e il turismo (2 aprile 1925–7 giugno 1952)). In the post-war period, in democratic Italy, Gasparotto, one of the founders of the progressive Democratic Labour party, was head of the Senate group for tourism, and condemned the new assaults on the landscape, both urban and natural.

First of all, he collected other complaints regarding the exploitation of sections of coastline with landscape value, as quarries for building material. For example, between 1947 and 1950 four quarries were opened on the Sorrento coast in the Gulf of Naples: one at Pozzano, two at Scutolo and a fourth at Marina di Alimuri. The complaint was made by a group of residents.

> We, residents of Sorrento, are particularly exasperated by the continual and systematic devastation being carried out daily by a few exploiters: without any control and without any conscience, they are dismantling rocks covered in vegetation, flowers and gorse, that were a joy to the eye of those who used to come to our country. Every day powerful mines explode the rocks, to the danger of passers-by and with immense damage to the landscape (Sesto San Giovanni. Fondazione Istituto per la storia dell'età contemporanea. Fondo Luigi Gasparotto. Unità 51 – Corrispondenza di Luigi Gasparotto e ritagli stampa sulla tutela del paesaggio e il turismo (2 aprile 1925–7 giugno 1952)).

In this case the rebuilding and furious urban expansion of the nearby city of Naples required building materials.

In another case, rebuilding was quickly changing the face of cities and in Milan, for example, advertising posters at the end of the 1940s were invading not only the entrances to motorways but also town centres, where they even crowded onto the Arco della Pace in Piazza Sempione, the Giureconsulti Palace in Piazza Mercanti, and the buildings of Piazza del Duomo and Piazza della Scala ("Contro i cartelli pubblicitari deturpatori," 1949).

However, the frenzy of reconstruction was also transforming those areas that were already tourist areas and that were waiting impatiently for the economic upturn, such as the pre-alpine lakes area near Switzerland, or the resorts in the Dolomites or, again, the Liguria region on the northern Tyrrhenian Sea up to the border with France, where a series of riviera resorts had already developed in the mid-19th century thanks to the climate and subsequently the sea-bathing. After the war, these lands were the first to receive influxes of foreign tourism: they were frequented by tourists able to travel by car and to overcome all the limitations on fuel distribution and on currency imposed by the war and still in force.

So, by collecting all the complaints from various associations, citizens' committees and local bodies, Luigi Gasparotto drew the government's attention on more than one occasion "to the need to safeguard scenic beauties" from the disfigurement being carried out with the indifference of the state authorities themselves.

Reconstructing some of these events may clarify the complexity of the dynamics in the places concerned.

In the 1920s and 1930s, certain boroughs already had building regulations in place, which they had had to adopt by law in order to be recognized as tourist resorts, so that they could apply a tourist tax and manage it independently – in accordance with article 20 of Royal Legislative Decree No. 765 of 15 April 1926, *Provisions for the protection and development of health care, accommodation and tourist resorts*. In those centres, state legislation safeguarding the cultural heritage and landscapes was thus reinforced by local authority regulations, which turned out to be more familiar to and shared by the population. If, for example, local authority committees illegally approved a project and sent it to the central offices of the State for final approval, the local communities had the tools to contest them, because they were well aware of the regulations and therefore of their own rights.

In 1949 Albissola Marina, a "resort for health care, accommodation and tourism" in Savona province, sent a petition with fourteen signatures to the Minister of Education, to stop the construction on the seafront of a thirteen-storey "skyscraper"-type building that would alter "the physiognomy of the town" and cause irreparable damage to the scenic beauties of the area. The

Supervisory Authority (*Soprintendenza*) and the Prefecture had stated that the new construction did not fall into an area bound by scenic beauty and therefore the project only needed to comply with local borough building regulations. The builder defended the construction with various arguments: he defined the building as a "modest house, which is not a skyscraper because it consists of 10 floors," that was built to offer apartments to families of workers: in addition, he made a personal appeal to Luigi Gasparotto in the name of a mutual past experience of antifascist exile in Switzerland in the war years. And he went even further, reminding him of the aid offered by his family to his son – later killed by the Germans – during the Resistance. However, the pressures brought to bear by designer and builder were not completely effective, and the project had to be reduced in size, thanks to the citizens' strong condemnation.

In 1950 in Santa Margherita Ligure, another health and seaside resort in Liguria, a group of citizens complained about a 5-storey building that would alter the landscape of the coastal resort. In this case the land was entailed and, although the Borough and the Supervisory Authority approved the project, the Ministry of Education forced the designer to build one storey less. However, the building also broke the local borough building regulations, which stated that the height of buildings could never exceed one-and-a-half times the width of the streets and squares. In this case the population complained that "such a colossal and contrasting building" would ruin "the delicate line of the sea" and, even though the building was built anyway, its height was scaled down.

Meanwhile, in those same years, the coastline began to be attacked by tourist businesses who were trying to meet an ever-larger demand for sea bathing.

Within just a few years from the end of the war, access to the sea for the citizens of Genoa became a real "taboo," as bathing concessions left only the "garbage corner" free. In fact, only the rocky areas remained available, such as the Nervi cliff, where the borough authorities tried to respect the natural beauty of the coast as far as possible, by siting the bathing beaches so as "not to harm the aesthetics of the coastline" (Bo, 1950).

In Spring 1949, a project was presented to the Borough and the Supervisory Authority for construction of a large bathing beach which would occupy a vast stretch of the cliff, with numerous bathing-huts and an artificial swimming pool set amidst the rocks. This project was immediately rejected by both the Supervisory Authority and the Port Authority. A second, slightly reduced project was then presented, providing for a line of around 30 masonry huts just behind the seafront promenade. This project was also rejected, and only after long examination was the building of just 10 masonry huts authorized. The Supervisory

Authority and Port Authority were very clear: no artificial works must be carried out on the cliff, under penalty of loss of the concession.

The various versions of the project were also assessed by the local tourist office, which included a commission of technicians made up of artists, engineers and architects. The office also vetoed the project for landscape reasons as well as for hygiene reasons, because a sewer emptied out underneath the facility and because the cliff had to be left to the resident population for bathing. In addition, the project did not provide for a solarium and access to the sea, which, it was feared, might be created with concrete being laid secretly at night.

The tourist office's opinion was backed by the provincial Tourist Board; both defended "a common good" and one for public enjoyment against private speculation.

In 1949–50, the Nervi cliff was defended by state public bodies but especially by the local population, by writers, journalists, the Senate parliamentary group, the provincial Tourist Board and by the local tourist office, which was prepared to make "every effort to defend the natural beauties of this country, a duty of care that falls to us not only logically, but because of precise legal references to the protection of natural beauties." The "absolute untouchability of the magnificent Nervi cliff," clearly limiting a private initiative defined as "anti-tourist," was thus declared (Sesto San Giovanni (Milano). Fondazione Istituto per la storia dell'età contemporanea. Fondo Luigi Gasparotto. Unità 51 – Corrispondenza di Luigi Gasparotto e ritagli stampa sulla tutela del paesaggio e il turismo (2 aprile 1925–7 giugno 1952).

Along with the coastal areas, the Dolomites region in the Alps was going through a process of transformation. From Selva Val Gardena, Osvaldo Pitscheider, a hotelier and chairman of the local tourist office, asked the parliamentary group for tourism to intercede with the Commission for Tourism in Rome, to urgently solve the "great problem of protecting the natural beauties and the landscape." Selva was offered as a winter sports centre and a health and tourist resort, but a large amount of construction had been going on since the post-war period. Pitscheider was not seeking a limitation on new buildings, but that they should have an architecture in tune with the place. Moreover, the chairman of the local tourist office also complained that the Borough's urban development plan had been compiled but had never been approved, and so everyone built wherever they wished and gave their building the features they liked, without taking into account the requirements dictated by the natural beauties of the area and of the Alto Adige in general (Sesto San Giovanni. Fondazione Istituto per la storia dell'età contemporanea. Fondo Luigi Gasparotto. Unità 51 – Corrispondenza di

Luigi Gasparotto e ritagli stampa sulla tutela del paesaggio e il turismo (2 aprile 1925–7 giugno 1952)).

5 Conclusions

In the early post-war years in Italy, defence of the cultural and natural heritage fuelled civil and political conflict and directly involved the local tourist institutions.

The events we have reconstructed show how directors and chairmen of the local tourist offices and the provincial Tourist Boards, from one end of Italy to the other and in resorts in different states of development, operated actively, analysing projects, condemning, supporting and sensitizing the press and local culture, through exhibitions, conferences and illustrated publications. This is what happened in Rimini, a major bathing resort on the Adriatic Sea, where "the various actors of the territory found a place for debate and cooperation" (Battilani, 2015, p. 607); it also happened in Siracusa in Sicily, where, having overcome the problems of reconstruction, in the 1950s the local tourist board "was characterised by its interventionism" (Criscione, 2014, p. 124), working at reorganizing services, advertising, restoring monuments, tending public parks, tidying the seafront and beaches, and at the same time carrying out an intense activity of reporting infringements and malfunctions, and monitoring facilities, prices and hygiene.

In any case, in the late 1940s and 1950s, the presence of those intermediate institutions in the territories was very important, because they were able to communicate with the central government bodies in Rome – which were undergoing a strenuous process of reorganization after fascism – and to interpret the local dynamics in the delicate reconstruction and recovery phase. In addition, within the context of all their activities they were definitely committed to protecting the cultural and environmental resources of the area they were responsible for.

That period, which is commonly interpreted as the start of the spontaneous phase of massification, should nevertheless be considered as a specific period in the history of tourism in Italy. The country was coming out of the long interlude of dictatorship, and communities, associations, local tourist institutions and the media immediately and enthusiastically resumed the public debate on themes of protecting natural resources and in particular landscapes, considered fundamental for tourism.

At the same time, however, in the newfound democratic climate, the world of tourism started to show deep internal divisions. On the one hand there were businesses that operated in tourism and spin-off industries; they had a long experience of management and were represented by the Chambers of Commerce – present in all provinces – that were very active and strenuously defended the local business community in the post-war period. On the other hand, there was a large bureaucratic apparatus created by the preceding fascist dictatorship made up of local tourist offices and by the provincial Tourist Boards, which acted as mixed intermediate bodies offering themselves as mediators in disputes and as local development regulators. After the fall of fascism, these peripheral bodies were accused of being part of a heavy bureaucratic system, inherited from the dictatorship to the extent that the Chambers of Commerce, and therefore private enterprise, asked that they be suppressed.

Throughout the whole of the 1950s, while the coasts were quickly being transformed to meet the needs of seaside tourism, the world of Italian tourism was being held back by the debate on governance, and the conflict between the private sector and peripheral tourist bodies was also reflected in the disagreements triggered around the use and the safeguarding of natural and environmental resources that were useful for tourism.

This essay therefore proposes an analysis of the local disputes that arose over environmental resources in the late 1940s and the 1950s in Italy, also in light of the conflict of an institutional nature that permeated the world of tourism.

The post-war centrist governments, now belonging to the western political sphere, needed to obtain and reinforce political consensus about these areas and oppose the left; they took up the challenge and achieved a mediation. We have a prime example of this in the documentation of the parliamentary group for tourism coordinated by Luigi Gasparotto. In September 1950, the Varazze section of the Christian Democratic party, in Liguria, upheld the struggle and the demands made by the population, the Borough and the local tourist office, for completion of the improvement works, already underway, on the seafront. These were to implement an urban development plan for the beach and oppose the interests of a few private citizens who had the backing of State agencies, which should be safeguarding the natural beauties of the town in order to increase tourism (Sesto San Giovanni. Fondazione Istituto per la storia dell'età contemporanea. Fondo Luigi Gasparotto. Unità 51 – Corrispondenza di Luigi Gasparotto e ritagli stampa sulla tutela del paesaggio e il turismo (2 aprile 1925-7 giugno 1952)). Nevertheless, the party in government, the Christian Democrats, could not take sides so openly; this is why the centrist governments made room for mediation with a series of reforms introduced in the 1950s: they guaranteed the survival of the peripheral bureaucratic system, but stripped it of its autonomy and subjected it to party interests, and in this way freed the work of private enterprises from excessively tight public control.

6 Sources

Amalfi (Salerno).
Archivio dell'Azienda autonoma.
1947, Amalfi, 8 settembre. Relazione del Presidente uscente ingegnere Francesco Ruggiero al Consiglio direttivo dell'Azienda.

Roma.
- Archivio centrale dello Stato.
Presidenza del Consiglio dei ministri. Atti di gabinetto.
1944/47–1945, fascicolo 1.6.1, 52178.
1944/47–1944, fascicolo 1.6.1, 17898.

Sesto San Giovanni (Milano).
- Fondazione Istituto per la storia dell'età contemporanea.
Fondo Luigi Gasparotto. Unità 51 – Corrispondenza di Luigi Gasparotto e ritagli stampa sulla tutela del paesaggio e il turismo (2 aprile 1925-7 giugno 1952).
https://patrimonio.archivio.senato.it

References

Battilani, P. (2015). "Dal turismo di massa al turismo relazionale: la Riviera romagnola." In M. Salvati & L. Sciolla (Eds.), *L'Italia e le sue regioni. L'età repubblicana. Territori* (pp. 603–610). Roma: Istituto della Enciclopedia italiana.

Berrino, A. (2011). *Storia del turismo in Italia*. Bologna: Il Mulino.

Berrino, A. (2012). *I trulli di Alberobello. Un secolo di tutela e di turismo*. Bologna: Il Mulino.

Bo (1950). "Il piccone distruttore ha già addentato la scogliera." *Corriere del popolo* (20 febbraio).

Camera di commercio, industria, artigianato di Genova (1947). *Problemi del turismo: memorie e note presentate al Primo congresso nazionale del turismo, Genova 15–19 maggio 1947*. Genova: Fratelli Pagano.

Ciccaglione, A. (1947). "Per la difesa delle spiagge del litorale italiano." In Camera di commercio, industria, artigianato di Genova. *Problemi del turismo: memorie e note presentate al Primo congresso nazionale del turismo, Genova 15–19 maggio 1947* (parte II, pp. 179–180). Genova: Fratelli Pagano.

"Contro i cartelli pubblicitari deturpatori" (1949). *Il volto della patria. Rivista dell'Associazione nazionale per i paesaggi ed i monumenti pittoreschi d'Italia. Bollettino di informazioni* (2 giugno), 4.

Corona, G. (2015). *Breve storia dell'ambiente in Italia*. Bologna: il Mulino.
Criscione, G. (2014). "Per una storia del turismo e del commercio." In S. Adorno (Ed.), *Storia di Siracusa. Economia, politica, società (1946-2000)* (pp. 101-142). Roma: Donzelli editore.
D'Angelo, L. (1999). "Gasparotto, Luigi." In *Dizionario biografico degli italiani*, volume 52 (pp. 494-495). Roma: Istituto della enciclopedia italiana.
Ente provinciale per il turismo di Salerno (1947). "La pianificazione delle località turistiche e della loro zona di influenza." In Camera di commercio, industria, artigianato di Genova. *Problemi del turismo: memorie e note presentate al Primo congresso nazionale del turismo, Genova 15-19 maggio 1947* (parte I, pp. 71-75). Genova: Fratelli Pagano.
Fiore, A. (2018). "La taglia su viandanti e forestieri: Napoli e Castellammare, 1840-1860." *Storia del turismo. Annale* (11), 9-23.
Francese, R. (1949). *Come fu svegliata dal sonno la meravigliosa Grotta d'Amalfi*. Salerno: Tipografia Jannone.
Gentile, E. (2007). *Fascismo di pietra*. Bari-Roma: Gius. Laterza & figli.
Groß, R., Knoll, M. & Scharf, K. (2020). "Where the Histories of the European Recovery Program (ERP)/Marshall Plan and European Tourism Meet. An Introduction." In R. Groß, M. Knoll & K. Scharf (Eds.). *Transformative Recovery? The European Recovery Program (ERP) / Marshall Plan in European Tourism* (pp. 7-32). Innsbruck: Innsbruck University Press.
Kavrečič, P. (2016). "Il litorale austriaco prima della Grande Guerra: il caso di Grado." *Storia del turismo. Annale* (10), 75-94.
Lanzani, A., Bolocan Goldstein, M. & Zanfi, F. (2015). "Della grande trasformazione del paesaggio." In M. Salvati & L. Sciolla (Eds.), *L'Italia e le sue regioni. L'età repubblicana. Territori* (pp. 291-312). Roma: Istituto della Enciclopedia italiana.
Large, D.C. (2019). *L'Europa alle terme. Una storia di intrighi, politica, arte e cura del corpo*. Torino: EDT.
Moranda, S. (2015). "The emergence of an environmental history of tourism." *Journal of Tourism History* (7, n. 3), 268-289.
Piccioni, L. (1994). *Il volto amato della Patria. Il primo movimento per la protezione della natura in Italia, 1880-1934*. Camerino: Università degli studi di Camerino.
Pivato, S. (2006). *Il Touring Club Italiano*. Bologna: Il Mulino.
Ruata, G. (1946a). "Abano Terme." *Turismo* (1, n. 2), 19-21.
Ruata, G. (1946b). "Le terme di Montecatini." *Turismo* (1, n. 3), 11-15.

Silvestri, G. (1946). "Problemi della ricostruzione: il volto delle nostre città." *Turismo* (1, n. 3), 24–26.

Volta, S. (1947). "Alcuni aspetti del problema delle spiagge nella Liguria occidentale." In Camera di commercio, industria, artigianato di Genova. *Problemi del turismo: memorie e note presentate al Primo congresso nazionale del turismo, Genova 15–19 maggio 1947* (parte II, pp. 210–213). Genova: Fratelli Pagano.

Zuelow, E.G.E. (2016). *A history of modern tourism*. London-New York: Palgrave and Macmillan.

Patrizia Battilani and Donatella Strangio

Mass tourism and social sustainability: insights from the Italian and French Coasts[1]

Abstract In the 1980s social sustainability found a clear conceptualization based on local community engagement and cultural diversity promotion. Recently the UN's World Tourist Organization (UNWTO, 2019), focusing on participatory processes, has recognized destination management organizations (DMOs) as key players in designing sustainable tourism. Despite its importance, tourism historians have only occasionally paid attention to the issue and DMOs have usually been studied more to assess their impact on promotion than on sustainability. This paper is a first attempt to put the social sustainability issue into an historical perspective by focusing on DMOs in France and Italy, from the interwar years to the 1970s. In Italy DMOs made their appearance in 1926 (they were called Azienda di soggiorno, cura e turismo) and soon started to design concerted tourism strategies at the municipal level involving local stakeholders, as the case study of Rimini and Riccione shows. They became a driver for social sustainability. Where tourism development was driven by the state, it was impossible to set up real DMOs and this also impacted social sustainability as the experience of Latina province reveals. Since the mid-1960s changes in mass tourism (media, advertising, transport) have pointed to the advantages of scaling strategies at the regional level. Italy and France have chosen different solutions, making the task of guaranteeing social sustainability more complex.

1 Social sustainability and tourism development: An introductory overview

In the beginning social sustainability was considered a sort of completion of development models, able to cope with the environmental challenge. The result is that even today its contours are blurred (Eizenberg & Jabareen, 2017). The environmental dimension was the first to capture attention after the environmental disasters that occurred in Europe and the US during the 1950s and 1960s (Fisher, 1953; Battilani & Fauri, 2019). It is no coincidence that the WWF was set up in 1961. At this time many international organizations introduced environmental protection into their charters and conventions. The first of these was the Council

---

1   This study forms part of the research project HAR2017-82679-C2-1-P, financed by the Ministry of Science and Innovation of the Government of Spain and the ERDF.

of Europe which in 1968 adopted a Water Charter and a Clean Air Charter based on the "polluter pays" principle. In 1972, on occasion of the Paris Summit of the Heads of the European Countries, the European Union started a process of reflection that led it in the 1990s to dedicate 4 articles of the TFEU (Treaty on the functioning of the European Union), numbers 11 and 191–193, to environmental policy. However, in those years the most important player on the international scene was the United Nations, which in 1972 established a new subsidiary body of its General Assembly, the United Nations Environment Program – UNEP, the forerunner of UNEO (UN Environment Organization 2005). In 1987 UNEO promoted the famous report "Our Common Future," which developed the notion of sustainable development in its three dimensions: economic, social and environmental. Written by the Bruntland Commission, it gave importance to some main goals of social sustainability. The first was equity between generations and within each generation, to be implemented in a transnational perspective through the concept of social justice within and amongst nations (UN World Commission on Environment and Development, 1987, p. 45). The second was the effective participation of local communities in decision-making processes as a way to help them articulate and effectively enforce their common interests (UN World Commission on Environment and Development, 1987, p. 45). The report also introduced a third concept, cultural diversity, with reference to specific communities, the so-called indigenous or tribal peoples. … "The isolation of many such people has meant the preservation of a traditional way of life in close harmony with the natural environment. … Growing interaction with the larger world is increasing the vulnerability of these groups, since they are often left out of the processes of economic development…. Many groups become dispossessed and marginalized, and their traditional practices disappear. They become the victims of what could be described as cultural extinction" (UN World Commission on Environment and Development, 1987, p. 98). According to this view such communities are the repositories of traditional knowledge and experience that link humanity with its ancient origins. Their disappearance would have been a loss for the entire world and would have also contributed to exacerbating the environmental issue, due to the fact that these communities have proved able to thrive in complex ecological systems such as rain forests or deserts. This permitted the recognition and protection of their traditional rights to land and the other resources that sustained their way of life. Over the last two decades, this vision has been strengthened through the new concept of cultural diversity promotion (UNESCO, 2001; UNESCO, 2005). In conclusion, the report identified three main components of social sustainability: social justice, participation in the decision-making process and preservation of minority cultures.

The report thus provided a forward-looking synthesis of the participatory experiments led by town planners in the 1960s in many European and American cities. See for instance the advocacy planning paradigm in the US, which introduced pluralism in urban planning in order to give a voice to the poorest parts of society in shaping the neighbourhoods where they lived (Davidoff, 1965). Equally important were the visionary experiments of some architects such as Giancarlo de Carlo who in 1969 tried to realize participatory urban planning in Rimini, Italy's capital of mass tourism. He believed that "architecture in the future will be characterized by an increasing participation of the user .... and contemporary architects must do everything possible to make architecture less and less the representation of its designers and more and more the representation of its users" (De Carlo, 1980; Marini, 2015).

In the tourism sector sustainability became part of the public debate in the 1980s when it began to influence tourism planning. This was partly due to the fact that the first generation of tourism plans which had taken shape during the 1950s were mostly concerned with physical planning: "resources were carefully analysed and later combined with a summary market survey to produce an attractive image of future tourism development, defining the necessary infrastructure and pointing out favourable locations for the various facilities required" (Baud-Bovy, 1982, p. 310). They were the expression of a top down approach in which local communities were considered mere executors of decisions taken by high profile technicians, large corporations or the central political power. According to Timothy only during the 1970s and 1980s did scholars move to "a more balanced form of planning that recognizes the need for greater community involvement and environmental sensitivity" (de Kadt, 1979; Krippendorf, 1982; Murphy, 1985; Getz, 1987; Inskeep, 1991; Timothy, 1999, p. 371). The unsatisfactory results of some physical planning stimulated a more decentralized approach which included social and environmental dimensions (Haywood, 1988; Long, 1993; Prentice, 1993; Simmons, 1994; Timothy, 1998). Town planners and geographers started to identify the main components of social sustainability. For instance, Timothy focused on the "involvement of community members in decision making, the participation of locals in the benefits of tourism, and finally education of locals about tourism" (Timothy, 1999, p. 372). Recently the concept of co-creation has further enriched the debate on participatory practices in tourism (Prahalad & Ramaswamy, 2004, XX).

The UNWTO recently published guidelines for Destination Management Organizations (DMOs) attributing to them a leading role in "promoting a greater engagement of the tourism sector, its industries, as well as policy and decision makers with sustainable development" (UNWTO, 2019, p.12). They are recognized as key players in strategic planning and consequently also in designing sustainable tourist destinations

Due to the fact that sustainability is an underrepresented topic in tourism history, DMOs' strategies have been studied more to assess their marketing policies than to understand their impact on social sustainability. The theoretical elaboration that came closest to the theme of sustainability was the distinction between the community and the corporate model. The former described a situation in which the design and promotion of tourist destinations derived from a coordination process involving the main local stakeholders, from entrepreneurs to public bodies. The latter, the corporate model, is a context in which tourism development and promotion are dominated by a single large company, which generally enjoys a leadership position in the provision of at least one of the basic services and guarantees the provision of all the others. In command economies, the state took this leading position. Usually the dichotomy between the corporate (or state) and community model is mirrored by the adoption of a top down or bottom up development process.

As you can see in Tab. 1, the community model has prevailed in Europe throughout the rise of mass tourism.

In this context DMOs could have played a pivotal role not only in designing and promoting tourist development but also in shaping social sustainability, at least at the local level. However, the extent of local stakeholder and community participation seems to have been more difficult when scaling up to the provincial or regional level.

In this paper we will analyse social sustainability in the context of tourism history by focusing on the ability of DMOs to involve local stakeholders in designing tourist destinations in three Italian and French regions. This will allow us to reassess the sustainability of mass tourism, drawing attention to the social dimension. At first glance, the latter seems incompatible with the definition of mass tourism usually provided by economists and sociologists. According to economists, in fact, the era of mass tourism began when large companies (tour operators, large hotel chains and charter flight companies) acquired a dominant role. By uniting innovation and standardization, these companies managed to exploit economies of scale and sell all-inclusive holiday packages at an affordable price even to the middle and popular classes. The keywords describing this new social phenomenon became escape, fun and transgression. More than any other experience, seaside stays embodied the new world contained in the imagery of the 5Ss (Sea, Sun, Sand, Sex, Spirit). Historians, in truth, have proposed a much more nuanced vision of mass tourism, pointing out the pivotal roles played by many local communities. However, until now no alternative narrative has

Tab. 1: Development model of some tourist destinations in Europe

Tourist destination	Top down versus Bottom up process	Community, state and corporate model	Timing	Scale	References
Emilia Romagna Coast (IT)	Bottom up	Community	1926–1970	Local	Battilani, 2009
Brindisi (IT)	Mixed model	State and community	1950–	Local	Caroppo, 2021
Emerald Coast (IT)	Top Down	Corporate	1960–1990	Provincial	Battilani, 2002
Flowers Coast (IT)	Bottom up	Community	1880–1980	Local	Zanini, 2012
Sunny Beach-Black Sea (BG)	Top down	State model		Regional	Valdéz & Spassov, 2007
Malaga (E)	Bottom up	Community	1900–1936	Local	Palau & Pellejero, 2020
San Sebastian (E)	Bottom up	Community	1880–1936	Local	Larrinaga Rodríguez, 2006
Ibiza (E)	Bottom up	Community	1931–2000	Local	Cardona* & Serra, 2014
Blackpool (UK)	Bottom up	Community		Local	Walton, 1978
Languedoc Roussillon (FR)	Mixed model	State and community	1963–1980	Regional	Racine, 1980

emerged. This research on DMOs' contribution to social sustainability aims to provide further insights into this issue.

2 The role played by tourist destination organizations in promoting social sustainability: A case study of the Italian Adriatic coast

Since the end of the 1920s Italian cities could count on local DMOs, called *Aziende di soggiorno cura e turismo,* to promote and design tourist services. In fact, the Royal decree law n. 765 of 15 April 1926 enabled tourist destinations to establish a DMO with a legal personality (articles 5, 8). DMOs weres governed by a committee consisting of one engineer and one doctor (designated by the provincial health council); two citizens appointed by the mayor and one representative of each of the following institutions or associations: Enit (the National Tourist Promotion Agency), the Italian Touring Club (the most important non-profit tourist association), the local hoteliers association; and the local shopkeepers association. The president was appointed by the prefect who was the provincial representative of the Ministry of the Interior. It is evident that the composition of the committee aimed to promote cooperation among the different stakeholders: private entrepreneurs, non-profit associations, the municipality, technicians and also residents. This is not very different from the quadruple helix model, based on cooperation between universities or research institutes, industry, NGO's (representing the various expressions of civil society) and local government, which the European Union is trying to foster at present (European Union, 2017).

The 1926 law designed for local DMOs was based on the same governance model as the National DMO (ENIT – Ente nazionale delle industrie turistiche-National agency for tourism industries), set up in 1919 (Syrjämaa, 1997; Battilani, 2020). ENIT was designed by one of the last democratic governments while the local DMO legislation took shape during the fascist dictatorship. The result was a reduction in the independence of DMOs, with presidents being appointed by prefects while the other committee members had to be approved by mayors (who at the time were no longer elected). Nonetheless, the DMO became the meeting place for different stakeholders who could interact with each other in designing tourism strategies even during the fascist period.

After the Second World War Italy's new democratic government decided not to change the governance of local DMOs and it was left intact until 1960. Therefore, at the municipal level tourism recovered thanks to the DMOs that continued to seek concerted strategies and contributed to negotiations among different stakeholders.

To shed light on this issue we analyse the functioning of some DMOs set up along the coast of Emilia Romagna (Riccione and Rimini) from the 1930s to the 1960s. In those decades this coastline experienced massive development of tourism as well as of the overall economy, designed through a bottom up process.

What is more, the whole community could enjoy the benefit of this inclusive model: landowners had the opportunity to invest in a new industry and also experienced an increase in their property values, fishermen had the opportunity to integrate the limited revenues from fishing with summer work at the beach as lifeguards or beach managers, the poorer social classes had the opportunity to find seasonal jobs in hotels, boarding houses or in managing beach services, and the urban middle class soon began to redevelop their homes to offer food and accommodation to less wealthy holidaymakers or to undertake artisanal production of beach equipment (Battilani & Bagnaresi, 2020).

Take for instance the functioning of the DMOs in Riccione set up 1928. In the beginning, the preparation of the Beach Regulations, which included the beach layout, prices, and the duties and rights of beach managers, was one of the most important activities as it strongly impacted the destination's image. The executive committee met repeatedly to discuss and partially change the regulations, before finding a shared solution. For instance, the hotelkeepers' association suggested a different price scheme. "Once the discussion is open, various amendments are proposed which, after appropriate discussion, are all accepted. But since the regulations cannot be considered definitive yet, their approval is postponed until the next session …." (Archive of Riccione DMO, 1932)

The Riccione DMO was also a place where requests, suggestions and controversies between barkeepers, the yacht club, hoteliers, beach managers and so on could be discussed and negotiated among (Archive of Riccione DMO, 9 October 1934; 7 May 1935).

This approach did not change after the Second World War, when beach erosion became one of the pressing issues. From 1962 on Riccione, like many other seaside destinations along the Adriatic Coast, experienced beach erosion. "The beach was everything" for that community. "Without its beach, today Riccione would be a small village without any resources" (Archive of Riccione DMO, 4 November 1964). In this context the DMO's executives met the beach managers association and other local stakeholders to discuss and choose the proper strategy (Archive of Riccione DMO, 18 November 1964).

The same approach was adopted in 1965 to stimulate renovation of the entire tourist destination when tourist arrivals from Germany started to decrease. The president of the DMO was quite aware of what was going on. Cheap flights to the Balearic Islands were attracting part of the northern German tourists, who until then had spent their holidays on the Adriatic Coast. Riccione needed to change its tourism model, increase and improve public green areas, promote its cuisine and traditional foods, address the problem of beach erosion and keep prices competitive. Again, the DMO organized a general meeting of all stakeholders to discuss the new promotional strategy, in collaboration with the municipality. At the end of the process the local community made available about 5,000 euros in extra funds at the time (in 2021 prices this would be 110,000 euros) to finance new promotional and advertising campaigns (Archive of Riccione AS, 1965).

Despite the good results of local DMOs in terms of tourism development and social sustainability, from the beginning of the 1960s it became evident that some problems could only be solved on a larger scale. As early as 1936 the fascist government had set up coordination agencies, the Enti provinciali del turismo (EPT), which organized meetings among the DMOs of the same province in order to define common strategies, when possible. This model based on national DMOs (ENIT), strong municipal DMOs and weak coordination agencies at the provincial level (the EPT), did not change until 1959, when the Ministry of Tourism, Sports and Entertainment was created. It took over the functions that the Constitution (Article 117) would have entrusted to region governments, which however had not yet been established. Even more important was the legislative reorganization of 1960 which increased the role of EPTs in terms of coordination and supervision of municipal DMOs, which continued to be responsible for tourism promotion and development but with a completely different executive committee. In fact, Presidential Decree no. 1042 of 27 August, focusing on DMO governance, stated that the DMO president should be appointed by the Minister for Tourism, after consulting with the prefect, while the Board of Directors should include a representative of the EPT, two employers and two workers belonging to the tourism sector, the mayor (or someone he delegated), a technician from the health council and three experts nominated by the prefect. The quadruple helix model of the interwar years was abandoned in favour of class representation (workers and employers). In addition, public intervention became stronger and non-profit associations were excluded. According to some historians a pyramidal structure was created with the ministry at the top, to which ninety EPTs and three hundred autonomous companies referred (Berrino, 2001). However, at least some DMOs continued to adhere to a participatory model aimed at involving local stakeholders, as in Riccione. Moreover, the president of Riccione's DMO continued to introduce himself as the top tourism manager in Riccione, thus emphasizing the central role of the DMO.

In another context, that of neighbouring Rimini, at the end of the 1960s the DMO began to lose part of its influence. This DMO had been set up in 1926 and since then had promoted and designed tourism at that seaside destination. After the Second World War, enhancement took the form of redevelopment of the beach and nearby areas and modernization and expansion of Miramare airport, in collaboration with the EPT (Archive Rimini AS, 1960, 1965 and 1970). Furthermore, like other DMOs Rimini's did its utmost to promote the image of the entire town and build long lasting relationships with foreign travel agencies. However, in 1967 the hoteliers' association decided to implement its own promotional campaign without even informing the local DMO, revealing a crisis in the organization as a mediator. The hoteliers association commissioned a British company to undertake a motivational survey in the UK. Unfortunately, the survey revealed evaluations that were anything but flattering for Rimini and, given that the project also included the free hospitality of some British journalists, these results were soon published in the Italian and foreign press: in July 1967 the Observer published an article in which it was stated that Rimini "was not a place where one could proudly say that they had been" (Observer 23 July), while in September a new article in the Travel Trade Gazette (15 September) stressed that "hotel rates were considered exaggerated compared to the quality of the service and the food offered." These negative judgments were soon taken up by the Italian press, where articles appeared claiming that "a holiday in Rimini is no longer stylish for the English" (Paese Sera 3 August). In conclusion, the initiative proved to be a boomerang for Rimini (Archive Rimini AS, Verbali del Cda 30/10/1967).

Starting in 1972 the model based on municipal DMOs was progressively dismantled, with the passing of tourism competences to regional governments (as provided for by the constitutional dictate) and the identification of a new organizational formula that abolished the local DMOs and EPTs, replacing them with Tourist Promotion Agencies (APT)s managed by public officials and with no relationship to local stakeholders. However, regions completed this transformation at different times. In some cases, local DMOs did not close down until the late 1980s.

Many problems and strategies now acquired a regional scale and DMOs stimulated more competition than collaboration between tourist destinations. To cope with the need for a more broad-based approach, Italy moved toward a region-based model of tourism promotion and planning. However, in doing so, the issue of social sustainability was completely forgotten. With the disappearance of the DMOs an important forum for involving stakeholders in tourism development also disappeared for at least two decades.

In summary we can say that the DMOs strongly contributed to making mass tourism socially sustainable, because they were a tool for meeting and representing the various stakeholders. In some territories they also stimulated consideration of crucial problems for the growth of tourism: in the cases of Rimini and Riccione they committed themselves to developing the airport and charter flights (in collaboration with the EPT), and led many initiatives aimed at raising stakeholder awareness of the need to build positive relationships with tour operators and upgrading hotels to the standards they required.

However, it must be emphasized that the DMOs' ability to contribute to tourism promotion was closely linked to the degree of development the sector had reached in those localities, given that they financed themselves through the tourist tax and beach management. On the other hand, the same law of 1926 provided for DMOs to be established in places where tourism was already an economic reality: their task was not to found tourist activities but to accompany their growth, precisely through the promotion and enhancement of the territory.

3 The case of the Pontine coast and Latina

Completely different is the case of the Pontine coast. The province of Latina is a geographical invention, created about seventy years ago following the reclamation of the Pontine marshes and the rise of other cities (called "new cities"), which became destinations for migration from other regions of Italy (Strangio, 2008, p. 33). The fascist regime, tracing new regional and provincial borders, suppressed the province of Caserta in 1927 so that the territory that composed it was attributed to the province of Naples, up to the Garigliano river, and to the province of Rome from the river up to Monte San Biagio; the territory extending from Monte San Biagio to the north already belonged to the province of Rome (Strangio, 2008, p. 33). In 1934 the ninety-third province of Italy, Littoria (which became Latina on 31 January 1945) was established, thus subtracting from the province of Rome the territory that stretched from Cisterna to the Garigliano river and the Ponzian islands to the province of Naples, which were thus assigned to the nascent territorial reality (Musci, 1996, p. 149). By decree of the Ministry for Press and Propaganda of 4 September 1936, the EPT of Latina province was established (Archive State of Latina, fondo *Ente provinciale per il turismo di Latina* (1937–1995) 510bb., 141 regg). Due to the fact that no local DMOs were operating there, it was more than a coordination body and was in charge of tourism development. However, the EPT was not able to stimulate stakeholder participation or promote social sustainability.

The current provincial structure is the consequence of political and administrative choices that have resulted in the separation of territories with common geographical or ethnic-cultural characteristics (Sottoriva, 2009a, pp. 29–49). As pointed out in the literature, an important turning point was the integral reclamation operation that took place from 1929 to 1939, although this had been preceded by a series of relevant attempts throughout history (Bevilacqua & Rossi-Doria, 1984; Rocci, 1995).

This area offers various aspects which when combined, make it an attractive tourist destination. The province's most important attractive features are natural, including the coast, islands, coastal lakes, Circeo National Park, thermal areas, mountains and hilly inland areas, with the coastal area being the most attractive aspect. Reports of the Historical Archive of the Bank of Italy (from 1946 to 1970) offer a complete overview of the macroeconomic reality of the province of Latina in those years (Historical Archive Bank of Italy, Banca d'Italia, Studi, pratiche, nn. corda 449, fasc.1 s.fasc.19; 452, fasc.1 s.fasc.40; 453 fasc.1 s.fasc.26; 456 fasc.1 s.fasc.2; 457 fasc.1 s.fasc.9; 460 fasc.1 s.fasc.4; 471 fasc.1 s.fasc.7; 473 fasc. 1 s.fsc.7).

The first report concerning the province of Latina dates back to 1946 and also included information on the annual economic trends of the major economic sectors of Rome and Frosinone. Since 1949, however, the reports on the province of Latina have been reported individually and in 1952 the tourism sector conquered its own well-defined space. From 1952 to 1956 there were complaints about the inadequacy of accommodation facilities. The most relevant type of tourism was occasional, consisting mainly of people transiting between Rome and Naples, and especially foreigners. At the time the main projects were road construction such as Via Flacca "to enhance the equally attractive coast of Sperlonga" and the development of some areas by ENIT (Historical Archive Bank of Italy, Banca d'Italia, Studi, pratt. n. 463, fasc. 1, s.fasc. 20). The year 1956 marked a positive change in tourism development in Latina as revealed "by a series of data that attest to how hotel equipment had improved quantitatively and qualitatively by a higher percentage than in previous years" (Historical Archive Bank of Italy, Banca d'Italia, Studi, pratt. n. 874, doc. 9). Up until 1964 tourism grew, on the basis of the area's natural beauty, but also due to the improved road network and the proximity to major centres of international interest. In particular, the opening of the Terracina-Sperlonga-Gaeta coastal road put Italian and foreign tourists into contact with the beauty of the local coastline. The government tried to promote and support tourism development in the province of Latina through the Cassa del Mezzogiorno.[2] This support was aimed at developing various sectors, for a total of 514,350 million lire and included a plan for the construction of new roads (Historical Archive Bank of Italy I, Banca d'Italia, Studi, pratt. n. 900, doc. 9).

Tourism development was based on the growth of second homes bought by neighbouring city dwellers for their summer holidays, "transit tourism" throughout the rest of the year, and week-end tourism as a consequence of mass

---

2 The "Cassa per il Mezzogiorno," "Cassa for Extraordinary Works of Public Interest in Southern Italy," was an Italian public institution, created in 1950 by the De Gasperi government, to finance industrial initiatives aimed at the economic development of southern Italy, in order to bridge the gap with northern Italy.

motorization and shortened work weeks. At the time, most tourists grouped in coastal destinations such as Terracina, Sperlonga, Formia, and Scauri. This first development ended in the mid-1960s. The decline recorded in the following years up until 1970 was due to changes in customer preferences, economic events and the opening of the Autostrada del Sole (Historical Archive Bank of Italy, Banca d'Italia, Studi, pratt. n. 939, doc. 5).

In a memoir of the Lido di Latina and the Pontine area dated 1969, reference is made to the development of tourism through collaboration between national and local stakeholders in implementing the plans of the "Cassa per il Mezzogiorno" (Sottoriva, 2009b, pp. 541–575).

At the EPT, studies and landscaping plans were also being prepared for the conservation and enhancement of the area's archaeological heritage and monuments. It is precisely the contrast between old and new that characterized the province: "young" cities arose alongside cities with a rich past, medieval castles and ancient Roman ruins. In addition to the variety of natural environments, the Circeo National Park (established in 1934) included finds and sites that testify to the presence of man from the most ancient to relatively more recent eras: prehistoric caves, ruins from the Roman era and 16th-century watchtowers of which Torre Paola is the only one that still retains its ancient structure. Finally, the myth of the sorceress Circe contributed to the fascination with the entire area of the park. While in the areas of the Lepini Mountains, north of the province, tourism was independent from that of the rest of the provincial territory, the area of the Ausoni and Aurunci mountains, on the other hand, was strictly dependent on the seaside; in fact, in this area the ends of the mountains stretch towards the coast, reaching in some cases right up to the sea (Corsetti & Nardi, 1994).

A report of the provincial authorities in the late 1960s highlighted the opportunity to implement policies which would enhance an area which until then had not been marred by mass tourism. The Report emphasized the fact that the landscapes of Turin and Viareggio, Rome, Torvaianica or Ostia were becoming more and more similar: the same crowds, the same cars, the same chaos, with the single addition of the "canned sea," as sociologists defined it (Sottoriva, 2009b, p. 549). On the contrary, Latina and its province were still far from a similar situation.

Interestingly, after the transfer of competences to the regional level, the provincial and local authorities were called upon to preserve the natural aspects and sustainability of those places, only marginally affected by mass tourism (given their physiognomy and history).

The same report which presented the features and opportunities for environmentally friendly tourism also aspired to greater participation of the local institutions and private stakeholders, eventually coordinated by the EPT of Latina in order to overcome individualistic and harmful positions (Sottoriva, 2009b, p. 550).

In a province founded artificially during the fascist regime, where tourism had been built up over time, it remained very difficult to foster stakeholder participation in designing shared tourism plans.

4 Social sustainability and the regional and national scale: The Languedoc Roussillion case study

In Italy throughout the 1950s and 1960s local DMOs operated in a context characterized by the absence of a national tourism policy. The Cassa per il Mezzogiorno tried to stimulate tourism in southern regions by focusing on the restoration of cultural heritage, strengthening road networks and building new hotels. However, the dialogue between the state (or its agencies), municipalities and local stakeholders remained occasional, as the Latina experience shows.

In contrast, the French context was quite different. In France, at the municipal level there were the Syndacat d'initiatives or Tourism Offices which were in some ways similar to the Italian DMOs. The Syndacat d'initiatives were non-profit associations set up by the local èlite to promote tourism since 1889, while the Tourism Offices were set up after the approval of a new law, on the 10 July 1964 (Manfredini, 2019). In the French case the composition of the executive committee tried to represent all the local stakeholders, while the mayor was appointed as president. The main difference with Italy was not at the local level, but in the different degree of state involvement in tourism planning.

At the end of the Second World War France designed a national plan as a tool for the reconstruction and development of the economy, with the establishment of the General Commissioner for the plan, whose first president was Jean Monnet. The economic strategy was based on five-year plans, in which major medium-term targets were indicated.

The years 1959–1975 can be considered the golden age of tourism's role in national planning with the launch of major programs of coastal development in Languedoc- Roussilon and the development of winter sports resorts in the Alps (Larique, 2006; Bodon, 2003; Bravard, 1987). The first experiences of tourism planning at the municipal level dated back to the immediate post-war period, when a completely new winter sports destination was built, Courcheval 1850 (Montaz, 2006).

Until the 1980s the central administration retained a pivotal role in planning. However, in 1982–3, following the laws on decentralization and the transfer of powers from state to regions, local authorities were given new powers, while planning was "regionalized" through the introduction of "planning contracts" signed by the state and regions.

In the golden age of state planning, the most famous case is that of Languedoc Roussillon, a region of 27,500 sq km and 200 km of coastline, located in the south of France, between the mouth of the Rhone and the Pyrenees, divided into 5 administrative departments (Aude, Gard, Hérault, Lozère and Pyrénées Orientales). The fate of tourism in this region was largely decided by massive planning initiated by the central state in 1963.

The region had always been poor economically, focused on agriculture and in particular on the production of poor quality wine, fishing and small port activities, so much so that it was affected by significant migratory flows towards the cities of central France. The situation had further deteriorated during the 1950s due to the serious crisis that had hit the wine sector. Overall, at the beginning of the 1960s it appeared as a backward region, where even tourism had failed to offer a significant source of income: in 1961 it received just 30,000 visitors a year. The comparison with the neighbouring French Riviera was as inevitable as it was disheartening.

The above-mentioned project was developed by the central state with the aim of diversifying the region's economy, creating new employment opportunities, responding to the growing European demand for seaside tourism, promoting tourism as an important surplus in the balance of payments, reducing pressure on neighbouring Côte d'Azur which was beginning to show signs of overcrowding, and eventually supporting social tourism (Klemm, 1992).

Specifically, through this plan, new localities were created (Port-Camargue, Grande Motte, Cap d'Agde, Gruissan, Port-Leucate, Port-Barcarès and Saint-Cyprien), for a total of 500,000 beds. Each locality was ensured its own distinct architectural style, and a balance was sought between building areas and green areas, both by creating protected areas and through reforestation (thousands of plants were distributed free of charge to residents, who were asked to collaborate). All the road and airport infrastructure necessary to guarantee an influx of tourists was built and part of the subdivision was destined for associations such as the trade unions, sports and educational organizations, to give shape to the structures to be reserved for social tourism (about 25 % of the land) (Racine, 1980; Murphy, 1985; Pearce, 1987). The results of the project, which lasted for twenty years, were undeniably positive as regards the promotion of tourism and the economic development of the area.

For the purposes of our reflection, what interests us is the type of tourism organization and the relationship between the public and private spheres that enabled the realization of a project of this magnitude and the impact it had in terms of social sustainability. It should be emphasized that in this case the main protagonist was the central state, which in 1963 launched the plan and specifically set up a public entity of an appropriate size, the Interministry Mission for tourism development of the Languedoc-Roussillon region. The head of the Mission was Pierre Racine, an experienced public official. This mission, made up of 20 officials and endowed with its own budget and powers, was entrusted with the task of organizing the skills necessary to carry out the project. In this way the central administration of the state took responsibility for the development of infrastructure, roads, ports and waterworks, for reforestation and the construction of sanitary and hygienic facilities, and finally for financial planning: he prepared and managed a plan with a regional dimension. In order to create a bridge between the central state and the local communities, committees were set up that included both representatives of the peripheral organs of the state and those of the local communities, who were entrusted with the task of preparing plans at the local level, where the overall approach of the regional plan was received and personalized. Finally, private individuals were entrusted with the construction of the actual settlements, giving them the building lots (even though the architects were hired directly by the government) (Scaramuzzi, 1993).

The Interministry mission was the tool adopted by the French state to overcome the limitations of municipal or provincial scales in designing tourism development. A complex organization was set up in order to give local community a voice. In conclusion this mission was an interesting attempt to ensure social sustainability in a large scale project.

In 1983, in order not to interrupt the dynamic created by the Interministry Mission of 1963, the Joint Union for Tourism Development was set up, headed by Racine himself, who remained in office until 1986, when he was replaced by Jacques Blanc, President of the Regional Council. Finally, in 1999, when the government decided to implement a more incisive coastal protection and enhancement policy, on the basis of the 1982 and 1983 laws on regional planning, the state and region signed a plan contract.

5 Conclusion

Mass tourism along the Mediterranean coast was characterized by a good level of social sustainability due to the permanent involvement of local stakeholders in designing tourism strategies. In Italy the municipal DMOs played a fundamental role in elaborating concerted strategies among different stakeholders, as witnessed by the case studies of Riccione and Rimini. They were a key player for social sustainability. More complex was the situation of regions in which tourism did not develop spontaneously and some kind of state intervention was required, as in the case of the Latina province and the Pontine coast. In this context the junction between the state and the municipality was the EPT, which however due to its type of governance could not really be a forum for mediation and concertation among different stakeholders. However, it tried to stimulate cooperation more than once.

During the 1970s and 1980s the need to scale up decisions and strategies to the regional level was a challenge for local stakeholders' involvement and consequently also for sustainability. The French solution of the Interministry Mission, despite the top down features, seems to have stimulated local stakeholders' collaboration more than the Italian regional model, where no forum was foreseen for dialogue between public and private stakeholders until the last decade of the century.

6 Sources

Riccione AS (Archive of the Azienda di soggiorno in Riccione)
- Fascicolo 1, 1932-Sistemazione dei servizi della spiaggia. Verbale di deliberazione-Regolamento per l'esercizio della Spiaggia (14/04/1932).
- Lettera al Signor Piccioni Pio inviata dal presidente del Comitato di cura, 9 ottobre 1934
- Richiesta del Club Nautico di lasciare senza cabine l'area di spiaggia di sua competenza per poter qui riporre i Dinghies, 7 maggio 1935.
- Delibere e Finanza. Anno 1964. Busta 307. Relazione dell'attività svolta dal consiglio di amministrazione dell'azienda di soggiorno dal 2 gennaio 1962 al 1964. Presentata il 4 novembre 1964.
- Delibere e Finanza Anno 1964. Busta Riunione del Consiglio direttivo dell'Associazione Bagnini presso l'AS.
- Delibere e Finanza 1965. Relazione sull'attività svolta nel 1965

Rimini AS (Archive of the Azienda di soggiorno in Rimini)
- verbali del cda 14/4/1960, 2/5/1960, 21/6/1960, 4/4/1970

Historical Archive Bank of Italy, Banca d'Italia, Studi, pratiche, nn. corda 449, fasc.1 s.fasc.19; 452, fasc.1 s.fasc.40; 453 fasc.1 s.fasc.26; 456 fasc.1 s.fasc.2; 457 fasc.1 s.fasc.9; 460 fasc.1 s.fasc.4; 463 fasc. 1 s.fasc.20; 471 fasc.1 s.fasc.7; 473 fasc. 1 s.fsc.7; 900 doc.9.

Archive State of Latina, fondo *Ente provinciale per il turismo di Latina* (1937-1995) 510bb., 141 regg

References

Battilani, P. (2002). "Rimini and Costa Smeralda: How Social Values Shape Recreational Sites." In S.C. Anderson and B. Tabb (Eds.), *Water, leisure and culture. European historical perspectives*. Oxford: Berg Publisher, pp. 209–221.

Battilani, P. (2009. "Rimini: a Mass Tourism Resort which based its Success on an Original Mix of Italian Style and Foreign Models." In L. Segreto, C. Manera & M. Pohl (Eds.), *Europe at the seaside. The economic history of mass tourism in the Mediterranean* (pp. 104–124). New York: Berghan Books.

Battilani, P. (2020). "Gli anni in cui tutto cambiò: il Turismo italiano fra il 1936 e il 1957," *TST –Transportes, Servicios Y Telecomunicaciones* (41), 103–133.

Battilani, P. & Bagnaresi, D. (2020). "La spiaggia come luogo di produzione e di consumo: dal modello informale ottocentesco a quello "taylorista" del periodo fra le due guerre." *Italia contemporanea* (294), 11–46.

Battilani, P. & Fauri, F. (2019). *Storia economica dell'Italia*. Bologna: Il Mulino

Baud-Bovy, M. (1982). "New concepts in planning for tourism and recreation." *Tourism Management* (3), 308–313.

Berrino, A. (Ed.) (2001). *A. Agosteo. Una vita nel turismo: ricordi di un funzionario ministeriale*, Napoli: Libreria Dante e Descartes.

Bevilacqua, P. & Rossi-Doria, M. (Eds.) (1984). *Le bonifiche in Italia dal '700 ad oggi*. Bari: Laterza.

Bodon, V. (2003). *La modernité au village: Tignes, Savines, Ubaye: la submersion de communes rurales au nom de l'intérêt général, 1920-1970*. Grenoble: Presses universitaires.

Bravard, Y. (1987). *Tignes, vie, mort et résurrection d'une communauté montagnarde*, St. Alban Leysse: Collection Trésors de la Savoie.

Cardona, J. R. & Serra, A. (2014). "Historia del turismo en Ibiza: Aplicación del Ciclo de Vida del Destino Turístico en un destino maduro del Mediterráneo," *Pasos* (12), 899–913.

Caroppo, E. (2021). "Sviluppo e limiti del turismo nel Mezzogiorno d'Italia negli anni della Ricostruzione: il caso della provincia di Brindisi in una prospettiva nazionale e internazionale." In Berrino, A. & Larrinaga, C. (Eds.), *Italia e Spagna nel turismo del secondo dopoguerra* (pp. 199–222). Milano: Franco Angeli.

Corsetti, L. & Nardi G. (1994). *Ricerche sull'ambiente naturale di Patrica e dei Monti Lepini*, Frosinone: Comune di Pratica

Davidoff, P. (1965). "Advocacy and Pluralism in Planning." *Journal of the American Institute of Planners* (31), 331–338.

de Carlo, G. (1980). "An Architecture of Participation." *Perspecta* (17), 74–79 de Kadt, E. (1979). "Social Planning for Tourism in the Developing Countries." *Annals of Tourism Research* (6), 36–48.

Eizenberg, E. & Jabareen, Y. (2017). "Social Sustainability: A New Conceptual Framework," *Sustainability*, (9). www.mdpi.com/journal/sustainability.

European Union (2017). *Using the quadruple helix approach to accelerate the transfer of research and innovation results to regional growth*. Bruxelles: Publication Office of the EU.

Fisher, J. L. (1953). "Natural Resources and Technological Change," *Land Economics* (29), 57–71

Getz, D. (1987). "Tourism Planning and Research: Traditions, Models and Futures." *Proceedings of the Australian Travel Workshop* (pp. 407–448). Bunbury, Western Australia: Australian Travel Workshop.

Haywood, K. M. (1988). "Responsible and Responsive Tourism Planning in the Community." *Tourism Management* (9), 105–118.

Klemm, M. (1992). "Sustainable tourism development." *Tourism Management* 13 (2), 169–180.

Krippendorf, J. (1982). "Towards New Tourism Policies: The Importance of Environmental and Sociocultural Factors." *Tourism Management* (3), 135–148.

Larique, B. (2006). "Les sports d'hiver en France: un développement conflictuel? Histoire d'une innovation touristique (1890– 1940)." *Flux* (63–64), 7–19.

Larrinaga Rodríguez, C. (2006). "Turismo y ordenación urbana en San Sebastián desde mediados del siglo XIX a 1936." In J. M. Beascoechea, M. González Portilla & P. Novo (Eds.), *La ciudad contemporánea, espacio y sociedad* (pp. 785–800). Bilbao: Universidad del País Vasco.

Long, V. H. (1993). "Techniques for Socially Sustainable Tourism Development: Lessons from Mexico." In J. G. Nelson, R. W. Butler & G. Wall (Eds.), *Tourism and Sustainable Development: Monitoring, Planning, Managing* (pp. 201–218). Waterloo, ON: Department of Geography, University of Waterloo.

Manfredini, J. (2019). "Faire d'une passion une profession: la place des syndicats d'initiative dans le tourisme français." *Mondes du Tourisme* (16), 1–25.

Marini, S. (Ed.) (2015). *Giancarlo de Carlo. L'architettura della partecipazione*. Macerata: Quodlibet.

Montaz, P. (2006). *Les Pionniers du tèlèskis*. Bresson: Impr. Cédigraphe

Murphy, P. E. (1985). *Tourism: A Community Approach*. London: Methuen.

Musci, L. (1996). "Il Lazio contemporaneo: regione definita, regione indefinibile." In G. Arnaldi & S. Bellezza *Atlante storico politico del Lazio. Regione*

Lazio-Assessorato alla Cultura, Coordinamento degli Istituti culturali del Lazio (pp. 125–166). Bari: Laterza.

Palau, S. & Pellejero Martínez, C. (2020). "Promoción turística y desarrollo geoeconómico 1900–1936: Málaga y Barcelona." *Ayer* (117), 189–220.

Pearce, D. (1987). *Tourism today: A geographical Analysis.* Essex: Longman.

Prahalad, C. K. & Ramaswamy V. (2004). "Co-creation experiences: The next practice in value creation." *Journal of Interactive Marketing* (18), 5–14

Prentice, R.C. (1993). "Community-Driven Tourism Planning and Residents| Preferences." *Tourism Management* (14), 218–227

Racine, P. (1980). *La Mission Impossibile: l'aménagement touristique du littoral Languedoc-Roussillon.* Montpellier: Midi Libre.

Rocci, G. R. (Ed.) (1995). *Pio VI. Le paludi pontine.* Terracina: Nuova poligrafica Gaeta.

Scaramuzzi, S. (1993). *Inventare i luoghi turistici: analisi di alcune esperienze significative.* Padova: Cedam.

Simmons, D. G. (1994). "Community Participation in Tourism Planning." *Tourism Management* (15), 98–108.

Sottoriva, G. (2009a). "La Provincia Divisa." In *Scritti di Turismo. Il mare, la collina, l'ambiente, la cultura, la gastronomia, varie* (pp. 29–49). Latina: APT.

Sottoriva, G. (2009b). "Memoria sul Lido di Latina e sul comprensorio turistico pontino." In *Scritti di Turismo. Il mare, la collina, l'ambiente, la cultura, la gastronomia, varie* (pp. 541–575). Latina: APT.

Strangio, D. (2008). *Turismo e sviluppo economico. Latina e il suo territorio.* Roma: Casa editrice La Sapienza.

Syrjämaa, T. (1997). *Visitez l'Italie. Italian state tourist propaganda abroad, 1919–1943: administrative structure and practical realization.* Turku: Annales Universitatis Turkuensis.

Timothy, D. J. (1999). "Participatory Planning. A View of Tourism in Indonesia." *Annals of Tourism Research* (26), 371–391.

UNESCO (2001). *Universal Declaration on Cultural Diversity.* http://portal.unesco.org/en/ev.php-URL_ID=13179&URL_DO=DO_TOPIC&URL_SECTION=201.html

UNESCO (2005). *Protection and Promotion of Cultural Diversity Expressions Convention.* https://en.unesco.org/creativity/convention

UNWTO (2019). *Guidelines for Institutional Strengthening of Destination Management Organizations* (DMOs). https://www.e-unwto.org/doi/book/10.18111/9789284420841

UN World Commission on Environment and Development (1987). *Our Common Future*. https://sustainabledevelopment.un.org/content/documents/5987our-common-future.pdf

Valdés, J. A. & Spassov, E. (2007). "El modelo de sol y playa llega a los países de Europa del este. Análisis de su impacto en la competitividad de destino turístico búlgaro Sunny Beach." In D. L. Olivares & J. E. Bigné Alcañiz (Eds.), *Turismo en los espacios litorales: sol, playa y turismo residencial* (pp. 207–220). Valencia: Tirant lo Blanch.

Walton, J. (1978). *The Blackpool Landlady: A social history*. Manchester: University Press

Zanini, A. (2012). *Un secolo di turismo in Liguria. Dinamiche, percorsi, attori*. Milano: FrancoAngeli

Bertram M. Gordon

"Sous les pavés, la plage": Sun, Sand, and Surf in French Tourism – The Evolution of an Image

Abstract "Sous les pavés, la plage," or "under the cobblestones, the beach," a phrase that gained popularity when scrawled on the walls of Paris during the May 1968 uprising, reflects the popularity of seaside and beach tourism in post-World War II France. France's acquisition of Nice and the extension in 1864 of its railways from Paris to that city attracted international communities of English, Americans, and Russians. These developments, together with the coming of paid vacations under the Popular Front government in the 1930s, helped fuel the development of the Côte d'Azur and popularize Sun, Sand, and Surf touring among the growing middle and working classes. The coming of inexpensive Renault 4CV and Citroën 2CV automobiles after the Second World War increased the flow, enhanced after 1950 by the creation of Club Med, which by the 1970s had become famous for its hedonistic "bouffer, bronzer, et baiser" [eat, tan, and make love] slogan. Popular films such as "Les Vacances de M. Hulot" [Mr Hulot's Holiday], starring Jacques Tati in 1953, helped put French beach resorts on the tourist map. This essay tells the story of the increased availability of France's seaside resorts to French and international tourists, women and men, from the second half of the 20th century to the present together with some of the problems brought by "overtourism."

1 Introduction: Sun, sand, and surf – A tourism imaginary

"*Sous les pavés, la plage*" or "Beneath the Cobblestones Stones, the Beach" became a famous image of vacations at the beach in the minds of the student revolutionaries pulling up the cobblestones of the streets of Paris in May 1968 (Calmeilles, 2018). The image of young people fleeing to the beaches after tearing up the streets catches the power of the popular imaginary of sun and beach in France as well as elsewhere around the world during the late twentieth and early twenty-first centuries. Indeed, the development of Sun, Sand, and Surf vacations, as with all forms of tourism, is linked to the evolution of tourism imaginaries, in the words of Rachid Amirou, the "totality of images and evocations tied to tourism." They embrace visions of "explorations, travels, pilgrimages, vacations, leisure, adventure, relationships to space, nomadism, wandering, and discovery, among others." (Amirou, 2012, pp. 25–26). (Unless otherwise indicated, all translations are by the author). Noel B. Salazar describes tourism imaginaries "as socially transmitted representational assemblages that interact with people's personal imaginings and are used as meaning-making and world-shaping devices" (Salazar, 2012, p. 864).

The story of "*Sous les pavés, la plage*" or the history of the images of sun, sand, and surf, is tied to the historical development of tourism to the beaches of France, largely if not exclusively along the Mediterranean and this interaction over time is the subject of this essay. Measuring tourism can be difficult because collected statistics cannot fully gauge the intricacies of how people spend their time, for example at a beach. (Gordon, 2011, p. 93). The most recently available United Nations World Tourism Organization statistics, however, prior to the 2020 Covid-19 pandemic, listed France in first place for international arrivals. Although suggestive, these figures can only hint at the numbers of vacationers to the beaches as they include all destinations in France and exclude domestic travel. (International Tourism Highlights, 2019). All available evidence, however, points to a dramatic increase in visits to the beaches for sand and sun from the early 20th century to the present together with changing imaginaries of the pleasures to be gained there.

2 *Bronzage* and tourism – A prehistory

The imagery of French coastal sites such as the Mediterranean shore or Biarritz as beach and sunshine tourism destinations has a long pre-history even if the concept of sun-tanning or *bronzage* is relatively recent. The etymological French dictionary *Trésor de la Langue Française* defines the noun *"bronzage"* primarily as a process of strengthening a metal object by giving it a bronze overlay but also references it as the "color brown taken by the skin after its exposure to the sun *[Couleur brune prise par la peau après son exposition au soleil]*" or sun-tanning, with the word listed as a derivative of *bronzer,* or to bronze, with its earliest usage in 1845 (Trésor, 1975). The reader is referred to the synonym *"hâle,"* or "tan" in English. The earliest use of the term *hâle* in the *Trésor* dates to Chrétien de Troyes in 1176 who wrote of the *"action du soleil qui brunit, dessèche, flétrit"* [action of the sun that turns brown, dries, withers] (Trésor, 1981). For use of the term *bronzer* as a verb meaning to suntan, one must await a citation from Françoise Sagan's, *Bonjour tristesse* in 1954 (Trésor, 1975). In English, the language of many of the travellers to the south of France who helped develop the Côte d'Azur as a tourism destination, the term "suntan" is traced to its appearance in the *Bury and Norwich Post,* a newspaper in Suffolk, England, that on 10 May 1809 carried an advertisement that read: "A cleanser and beautifyer of the skin...may be used with the utmost safety on complexions the most fair and delicate for the removal of sun-tan, freckles, &c" (Oxford English Dictionary, 2021).

None of these definitions connote a positive view of sun-tanning or *bronzage*. Instead, sun-tanned skin was considered lower class in the United States and Europe, associated with those who laboured outdoors. Lighter coloured skin was believed to reflect refinement, especially for women who used parasols to protect their skin from the sun and maintain pale complexions (McKenna, 2009, p. 151). This attitude began to shift early in the 20th century with a growing recognition of the therapeutic benefits of sunlight discovered in the curing of rickets and the tuberculosis skin lesions of lupus vulgaris, among other illnesses. The increased desirability of sun-tanning as healthful among the European leisured classes was evidenced in an article describing a visit of the Prince of Wales in September 1913 to the German town of Sigmaringen, whose castle was described in the *Times* of London as possessing "many delightful terraces which could be adapted for sunbathing" (Sun-tanning, 1913). The more popular aversion to sun-tanning changed only in the 20th century, often attributed to its being popularized by the French fashion designer Coco Chanel after she was sun-tanned on a cruise off the Riviera coast in the 1920s (Wilkinson, 2012). Coco Chanel's star power appears to have been reinforced in France by the popularity of the American expatriate Josephine Baker, whose caramel colouring has also been said to have enhanced the vogue of sunbathing there.

The vogue for sunbathing may have been relatively recent but there is a much longer history of travel and tourism in France, enhanced with the publication of books following the invention of moveable type by Johannes Gutenberg in the middle of the 15th century. Increased centralization of the French state, accompanied by the construction of better roads and canals and increased domestic security, especially after the end of the religious wars in 1589, all facilitated travel. *La guide des chemins de France* [The Guide to the Roads of France] by Charles Estienne, published in 1552, outlined 283 itineraries in France and described local foods in a manner that Antoni Maczak later called "the prototypes of dishes later given star ratings in the Michelin guides" (Maczak, 1995, p. 25). Those who wrote accounts of their travels included François Rabelais (1494–1553) and Michel de Montaigne (1553–1592).

With a long coastline, France has beaches spreading along the English Channel from the Normandy shore through Brittany and the Bay of Biscay but the best known, those with the most widespread imagery, or what in French might be called *"rayonnement,"* of sun, sand, and surf are, arguably, those in the south that front the Mediterranean, notably the Côte d'Azur, also known as the French Riviera. Through the 18th and 19th centuries the image of the Côte d'Azur as a desirable destination was enhanced, if not created, by travellers from beyond the borders of France, often from England.

3 Early beach tourism in France

The "discovery" of the south of France by travellers from the north was a gradual process, connected with religious pilgrims on their way to sites such as Santiago de Compostela and others, largely young gentlemen and their tutors from England, passing through France as they made their way on the "Grand Tour" to sites of classical antiquity in Italy. Addressing the tourism patterns of the northern European aristocracy, Nelson Graburn described the process by which:

> the ruling families and the very wealthy began to leave their palaces for recreational and health reasons on a regular, yearly basis. Not since Roman times had this been done on such a massive scale … . Starting in the eighteenth century and becoming the mode in the nineteenth century, luxurious rivieras were built along the Mediterranean and Adriatic shores to house the royalty and idle rich from the nations of Northern and Eastern Europe. (Graburn, 1989, p. 30)

The pattern of northern aristocrats seeking "warmer abodes" in the south, he added, led to the creation of Monte Carlo and other similar sites. By the beginning of the 20th century, the building of luxury steamships made it possible for affluent Americans to join the northern European influx (Graburn, 1989, p. 30). Its connection to the warm winter climate was such that in a book about the "invention" of the Côte d'Azur, Marc Boyer added the subtitle "Winter in the South" (Boyer, 2002).

In their quest of the healthful benefits of the warmer south, visitors from the north began spending extended periods along the Côte d'Azur. Looking for accommodations, they sought secondary residences in the mid-18th century. In what might be seen as an earlier form of medical tourism, albeit without the surgery often considered as part of today's medical tourism (Berger, 2013, p. 25), these visitors, known in French as *villégiateurs,* or vacationers, helped give coastal areas such as Dieppe in Normandy and the Côte d'Azur an allure or mystique, or to put it in tourism studies terms, tourism "imaginaries" whose evolution continues to the present. Seaside tourism intensified during the 19th century with coastal resorts as health spas, at first visited with medical prescriptions. They were sought for the sun, the moderate sea climates, and the pleasurable activities at the resorts (Peyroutet, 1995, p. 100).

Boulogne and Dieppe, in Normandy, developed as resort areas during the Bourbon Restoration years that followed the defeat of Napoleon in 1815, with the royal court moving to Dieppe every July. Writing in the mid-20th century, Roger Babulle observed that being a *villégiateur* was uncommon prior to the French Revolution of 1789 because etiquette required the aristocrats to be present at the Versailles court except for visits to their own ancestral château or to accompany the court on a visit to "the waters." In addition, the sea was generally seen as something terrifying, to be avoided. From a medical viewpoint, bathing in the sea was considered a remedy for dementia. In 1820, the Duchess of Berry, the daughter-in-law of the restoration King Charles X, came to bathe in the waters at Dieppe, provoking a scandal that lasted only briefly, as the aristocrats, together with the rising *haute bourgeoisie*, were looking for vacation possibilities (Babulle, 1954, p. 67). Other beaches in Normandy, including Luc-de-Mer and Trouville, became fashionable during the 1840s (Rauch, 1996, p. 15).

Beaches began to increase in popularity in 1838 when Biarritz, on the Basque coast, was visited by Eugénie de Montijo, then a twelve-year old Spanish princess exiled along with other royalist Carlists. No longer able to visit San Sebastián, their preferred resort in Spain, they travelled instead up the coast to nearby Biarritz, where Eugénie and her family continued to vacation. After marrying Napoleon III in 1853, she began building her own palace in 1855. Known at first as the Villa Eugénie, and now as the Hôtel du Palais, it is located on what is now Biarritz's main seaside promenade. Guests of Napoleon and Eugénie included Queen Victoria and Edward VII of England (Babulle, 1954, p. 68). Aided by the increased accessibility facilitated by the Paris-Hendaye railway line in 1860, Biarritz was advertised as *"la Reine des plages et la plage des Rois"* [the queen of beaches and the beach of kings] (Rauch, 1996, p. 15). By 1900 it had become, in the words of André Rauch, writing about the history of vacations in France, "the luxury resort on the Ocean." In addition to visitors from other parts of France, Biarritz and other seaside sites attracted a European international aristocratic clientele, who as Rauch wrote, saw one another more as cousins than English, German, or French (Rauch, 1996, pp. 15–16).

Perhaps no one did more to put Nice and the Riviera on the tourist map than Tobias Smollett, whose *Travels in France and Italy*, published in 1766 extolled the then small town for having cured his asthma (Kanigel, 2002, p. 20). Seeking relief and travelling in France in 1763, Smollett had been advised by a chance acquaintance at an inn where he was staying while visiting Boulogne to visit Nice for the benefits of its climate. His discovery was described in 1973 by Nan Gillespie, who wrote in the *New York Times* that southern France was then becoming known as a place to help those with lung ailments. Gillespie added:

> Hence it was that, on a day in May of the year 1764 he had himself carried to the beach in a sedan chair, waded into the water and began to splash around. On this day, which unfortunately he does not date for us, Nice found her vocation for the future. (Gillespie, 1973)

Smollett's book, Gillespie adds, aroused the interest of members of the British royal family, helping to make Nice, then part of the Kingdom of Piedmont-Sardinia, a fashionable resort. King George III's brother, the Duke of Gloucester, visited in 1770, followed by other members of the British aristocracy who came to spend their winters there. Eventually, a British colony developed on the right bank of the Paillon River that ran through the city but was later covered over. British visitors, following the lead of Smollett, began spending their winters in southern France during the late 18th and 19th centuries and for many of them Nice became a site for second residences.

Cannes was "discovered" in 1834 by Lord Henry Peter Brougham (1778–1868), an English aristocrat, during a forced stay there due to a quarantine order preventing him from traveling to Nice. He had an Italianate villa built for himself and other English aristocrats followed (Cannes, 1995). By the late 19th century, Cannes had become the site of many villas and almost fifty hotels. Nice possessed 31 hotels in 1861 and 54 in 1877 (Schor, 2010). In Cannes, flower shows and regattas entertained the "fashionable gentry from Queen Victoria's court" (Cannes, 1995).

Interest was enhanced following the 1859 Austro-Sardinian War when the French acquired Savoy from Piedmont. The new acquisition enabled the French to develop the areas around Nice, Cannes, and Saint-Tropez, which together put the Riviera on the tourist map. Rail lines, extended from Paris to Nice in 1864, helped fuel the subsequent development of the Côte d'Azur, linking it with London and Paris and making the trips both faster and safer. The *Compagnie des chemins de fer de Paris à Lyon et à la Méditerranée* (PLM), established between 1858 and 1862, eventually connected Paris to the Côte d'Azur by way of Dijon, Lyon, and Marseille. Whereas it took Smollett some two weeks to travel from Paris to Nice by coach, the trip by train was eventually shortened to less than a day. At the end of the 19th century, steamships reduced the time for the Atlantic crossing to six days (Levenstein, 1998, p. 125).

The first casino in Nice was inaugurated in 1867 on the Promenade des Anglais. In 1887, Stéphen Liégeard, a lawyer, writer, and parliamentary deputy in France, published *La Côte d'Azur*, a book that popularized the term (Schor, 2010). The Côte d'Azur emerged as a tourist destination with British aristocrats traveling to Hyères and Giens, in the Var, often for warmer weather during the winter. They helped establish winter resorts on the Mediterranean, thereby creating the winter tourist "season" (Rauch, 1996, p. 20; Gordon 2003, pp. 207 and 209).

Many of the affluent visitors acquired secondary residences, which better enabled them to enjoy their stays along the coast because, as Louis Bonnard noted in 1927, the quality of hotels in mid-19th century France as a whole was low. He cited the historian Hippolyte Taine, who as an examiner for the military École de Saint-Cyr, had to travel all over France, as describing the poor quality of the hotels in the Côte d'Azur town of Berre in 1865. Writing in his *Carnets de voyage*, during the 1860s, Taine complained: "The main hotel is on the beach. It is a large barracks from the last century, cracked, abandoned, filthy as a Spanish posada." In the preface to his *Itinéraire Générale de la France*, published in 1861, Adolphe Joanne, the author of travel guidebooks to France, worried about the future of tourism in France without an improvement in the quality and accessibility of its hotels. The situation improved toward the end of the century, Bonnard wrote, with increased tourism stimulated by the coming of the bicycle, the automobile, and the creation of organizations that promoted tourism, including the Touring-Club de France and the Automobile-Club de France, founded in 1890 and 1895, respectively (Bonnard, 1927, pp. 139, 147–148, and 152–154).

Writers extolling the Riviera during the middle of the 19th century included Alexandre Dumas, George Sand, and Alphonse Karr. Karr, a journalist, relocated to Saint-Raphaël in 1865, which also began to attract additional literary figures. Karr, whose reminiscences, *Livre de bord,* were published in 1879–1880, is now considered "the discoverer of Saint-Raphaël." (Karr Alphonse, 2021). Foreign communities included a Russian colony that began in 1856 when the Tsarina Alexandra visited for the season. "Maximilian II of Bavaria gave an impetus to the German 'invasion' of the eighteen-sixties" (Gillespie, 1973). Prominent individuals who helped promote tourism to the Mediterranean included the Prince of Wales in Monaco, Napoleon III in Nice, and Queen Victoria in Hyères (Lanquar, 1995, p. 36). The publication of the book *Winter and Spring on the Shores of the Mediterranean* by the English doctor James Henry Bennet (1816–1891) in 1861 also helped popularize the French Riviera as a winter holiday destination. Queen Victoria's decision to visit Menton was influenced by Bennet's book, which was soon published in a German-language edition in 1863 and also in the United States in 1870, leading to increased traffic from those countries as well (Nelson, 2007, pp. 11–13). The British influence made itself felt in numerous ways in France in general where, for example, Pau in the southwest was the scene of the first golf course in 1841, the first steeplechase in 1856, and the earliest fox hunting during the mid-19th century. The first tennis courts in France were established in Cannes.

4 Seeking the Sun – The 20th century

The beginning of the 20th century saw what Marc Boyer called the "invention" of the "Mediterranean summer," to which, he added, must be added its "cousin," the "tropical sun." These concepts, he emphasized, were born of the elites, especially the wealthy Americans who also "literally invented" Florida, an image of coconut trees, long and sandy beaches, and warm water in winter. It was these Americans, some with residences in Montparnasse, who on the eve of the First World War, propagated the imagery of traveling to the Mediterranean coast for sun, sand, and beach in the summer. The relaxed behaviour of these nouveaux-riche Americans together with the casual atmosphere of the "lost generation," created the Juan-les-Pins of Mistinguett and Charles Boyer (Boyer, 1999, p. 28).

The early 20th century saw an enhanced focus on the health advantages of exposure to the sun. A French poet, Théo Varlet, described by Jean-Didier Urbain as "a Robinson of the Île du Levant," one of the Îles d'Hyères, French islands off the Riviera coast, wrote in 1905:

> I had never before felt so clearly the marvelous attraction, the grandeur of the sun's kiss. I had yet to experience in my flesh, to live for myself, what were still just literary notions: the sea as creator of primordial life, the sun as father of life on our planet. (Urbain, 2003, p. 145)

The extension of rail lines continued to contribute to the popularity of the coast and its beaches in Normandy as well, whose popularity was augmented by a series of paintings that depicted them between 1863 and 1865. Postcards depicted the beach at Deauville, which now acquired its "season." An article published in the American periodical *Harper's Weekly* in 1890 reflected the increased interest in the beaches of Normandy, now accessible by train from Paris and other metropolitan centres (Gordon, 2018, pp. 30 and 238).

The destruction of much of the European aristocracy occasioned by the First World War and its aftermath brought significant post-war changes to the Côte d'Azur. Russia's revolutions of 1917 and economic problems in post-war Germany cut the numbers of affluent visitors from those countries. The aristocratic families that had maintained secondary residences there were replaced by the wealthy Americans discussed above by Boyer. Winter vacation sites for what Nelson Graburn called the "elitist 'international set' " were turned into "summer pleasure resorts" (Graburn, 1989, p. 30).

The Mediterranean took off as a destination in the 20th century, as it shifted from a winter resort for the elites to a more popular destination increasingly available to broader sections of the middle classes, including women. Films and the popular press of the mid-20th century helped promote images of it as a tourist delight and it shifted from a resort for the aristocratic *villégiateur* to more of a destination for tourism in the more modern sense of shorter stays in hotels rather than longer-term stays in secondary residences to cite a distinction made by Florence Deprest (Deprest, 1997, p. 12). While Mediterranean seaside tourism was evolving from longer to shorter stays during the 20th century, the numbers of visitors increased. Seaside tourism was showing significant growth as defined by Valene L. Smith in 1989: "a tourist is a temporarily leisured person who voluntarily visits a place away from home for the purpose of experiencing a change" (Smith, 1989, p. 1). More recently, Jean-Michel Hoerner and Catherine Sicart noted that there are many kinds of tourists and these include the nearly 80 % of them who end up in resorts (Hoerner and Sicart, 2003, p. 17).

In addition to the fashion shift from pale skin to sun-tanning, the immediate post-World War I years brought an increased use of automobiles, making beaches in all of France increasingly accessible to domestic as well as international visitors. The production of relatively inexpensive cameras added another dimension to the experiences of tourists along the beaches of the Mediterranean and elsewhere. Improved hygienic measures made it increasingly safe to visit the Mediterranean during the summer. As a consequence of all these changes, the interwar years saw the start of a shift from winter to summer tourism, as indicated in an article "Winter cruising in a summer sea," published in 1923 in *Travel Magazine* (Travel Magazine, 1923, pp. 28–32).

The enhanced accessibility and popularization of beaches in France, together with the changed attitudes toward sun-tanning found expression in interwar films and literature, notably the French novel and film, "Le Train Bleu" and the novel *Tender Is the Night*, the latter published by F. Scott Fitzgerald in 1934. "Le Train Bleu," released in 1927, was based on the novel by Maurice Dekobra, *La Madonne des Sleepings,* a story of prostitution and romance on the Paris-Marseille night sleeper train, which had begun service at the beginning of the 20th century. Dekobra's novel, however, turned it into a famous line and focused attention on the Côte d'Azur. The novel was said to be on every French railway bookstall during the interwar years. The semi-autobiographical story of wealthy American expatriate life on the Riviera between the wars, Fitzgerald's novel is replete with accounts of days spent with sports and swimming, on yachts, and in cafés. At the time, American women possessed voting rights and legal rights to their inheritances, and more of them could afford vacation travel unlike many of their counterparts in France and Italy (Gordon, 2003, pp. 213–214). *Tender Is the Night* was made into a film in 1962 in Hollywood, featuring Jennifer Jones and Jason Robards, two stars of the cinema.

The increased accessibility of the regions helped sun tanning surge in popularity between the wars. With automobiles largely replacing trains during the 1930s, and the passage of France's paid vacation law in 1936, tourism to the coasts increased dramatically. The coast drew some three million vacationers in 1936, a figure which grew to 16 million in 1978, and more than 30 million in 1995. This meant 317 vacationers for every one hundred permanent inhabitants along the Languedoc-Roussillon coastline, with similar figures elsewhere, ultimately leading to conservation efforts to protect the sites from what today would be called "overtourism" (Peyroutet, 1995, p. 102).

The growing popularity of Deauville on the Normandy coast as a tourist attraction was reflected in the Jean Delannoy film "Paris Deauville" in 1934 (Gordon, 2018, p. 39). Four films, three in German, addressed Nice: "*À propos de Nice*," which focused on the bourgeois tourists in that city (Williams, 1992, p. 217) and "*Nizza*" both in 1930, "*Flucht nach Nizza*" [Flight to Nice] in 1933, and "*Blumen aus Nizza*" [Flowers from Nice] in 1936. In consequence, the French Riviera has been described as "probably the most famous piece of coastline in the world" with its heyday in the 1930s "in its modern sense- that is, as a place to visit for its long glorious summer rather than, as Queen Victoria did, for winter warmth" (de Courcy, 2019, p. 1).

For industrial workers labouring indoors rather than outdoors, a suntan acquired a new meaning. Whereas previously, the skin of workers had been bronzed by their agricultural labour in the fields, leaving a lighter skin tone with higher social prestige, now, as Marc Boyer noted, the shift in their workspaces to factories gave outdoor tanning [*bronzage*] a special status, the look of returning to work after a vacation at the sea or in the mountains. They now looked to Saint-Tropez, Saint-Paul de Vence, and Vallauris on the Côte d'Azur (Boyer, 1996, pp. 116–117). On the Côte d'Azur in 1935, Dimitri Philipoff, a Russian refugee, with some friends, created the Club de l'Ours blanc, a non-profit organization designed to make vacations affordable to people of modest means. The Club emphasized sports and swimming in the Mediterranean and, of equal if not greater importance, cavorting in skimpy swim suits. It became the forerunner of the Club Méditerranée (Club Med) (Ehrenberg, 1990, p. 118).

Describing the influx of noted personalities, such as the Duke and Duchess of Windsor, following the Duke's abdication of the British throne in 1936, Anne de Courcy wrote:

> It was fitting that they had come to the Riviera, the glamorous, golden, sun-filled coastline famous for uninhibited enjoyment and where nobody enquired too deeply into your past. Here, every year, the rich, the famous, the beautiful and the eccentric gathered

to swim, gamble and soak up the sun in a hedonistic lifestyle that then seemed never-ending. (de Courcy, 2019, p. 3)

Following France's military defeat by Germany in 1940, many aspects of life there changed dramatically but the allure acquired by the Côte d'Azur remained. Tourism picked up in Nice in the spring of 1941 with some of the larger hotels run by German or Italian interests. Moneyed clientele in August 1943 were said to be enjoying expensive wines in the Savoy spa hotels (Gordon, 2018, p. 81). The immediate post-war years were difficult as many of the beach resorts had lost clientele due to the 1929 economic depression, which had taken hold in France during the mid-1930s. L. R. Blanchard, an American journalist on tour in the early spring of 1946 with several other American journalists, commented on the presence of a few "fashionable visitors" in Nice "who fight their way into accommodations of some sort" (Blanchard, 1946, p. 46). Recovery, however, was relatively swift. In Nice, for example, some 340,000 visitors in 1948 expanded to 540,000 by 1953 (Schor, 2010).

5 Sun-tanning and a tourism boom – Post World War II

The ethos of sun and sand was ushered in during the post-war period as tanning became very much in demand, intensifying and, to a degree, shifting demand for and the images of the Mediterranean beaches, starting first in France and Spain, then spreading to Greece, Italy, and Yugoslavia, and more recently to North Africa and Turkey (Urry, 2000, pp. 37–38). Inexpensive automobiles, such as the 4CV Renault and 2CV Citroën models, and more air travel both fed the growth in tourism to the beaches. The first package holiday by air in the U.K. brought visitors to Corsica in 1950 (Shaw and Williams, 1994, pp. 186–188). Cruise lines brought more tourists to the coastal sites as well. Nice became the second most popular destination for air travel in France.

Fashion styles also contributed to the imagery of the Côte d'Azur with the introduction in 1946 of the "bikini," defined as a "Woman's brief two-piece swimsuit with tiny bra top and brief pants cut below navel." While an earlier version covered the navel, a subsequent design, later in the same year, by Louis Réard, a clothing designer in Paris, exposed it. Réard named his swimsuit for the Bikini Atoll, where the first public test of a nuclear bomb had taken place four days previously. The "shock value" of its name and its configuration created "a sensation on Riviera beaches." The bikini grew even smaller by the 1980s (Calasibetta, 1988, p. 563).

With France recovering from the effects of the Second World War and occupation, some sought the economic benefits to be gained from the encouragement of American tourists to visit, whereas others, in particular Communists, argued against the invasion of American influence and power. Virgile Barrel, a Communist who served on the Nice Municipal Council, supported the encouragement of American tourism, whose dollars, he argued, would help the working class, thereby facilitating their own tourism and enjoyment of French sites later (Endy, 2004, p. 72). Tourism along the coast in Nice had developed to the point where the author of a guidebook for Americans driving through Europe blamed the noise of automobile traffic for the Côte d'Azur having lost "some of its charm" (Meyer, 1953, p. 90).

With the image of the Côte d'Azur firmly established by the 1950s, Alfred Hitchcock's movie "To Catch a Thief," starring Grace Kelly and Cary Grant in 1955 was set in the high American lifestyle of the Riviera, described by Bosley Crowther, a film critic for the *New York Times* as focusing on "the jewel thief whom Mr. Grant is stalking through the lush gambling-rooms and gilded chambers of French Riviera villas, casinos and hotels." Discussing Hitchcock's direction of the film, Crowther added:

> Most of his visual surprises are gotten this time with scenery—with the fantastic, spectacular vistas along the breath-taking Cote d'Azur. As no one has ever done before him, Mr. Hitchcock has used that famous coast to form a pictorial backdrop that fairly yanks your eyes out of your head. (Crowther, 1955, p. 14)

Most of the travellers to the beaches, however, were European rather than Americans. Citing French government studies, Christopher Endy writes that Americans, "along with other long-distance tourists from Canada, South America, and, later, Asia, more often cited monuments and museums as France's cultural attractions." The Côte d'Azur, however, Endy added, was an exception. Americans visiting France who ventured beyond Paris toured largely in the Loire Valley and the Riviera, "where they joined more numerous European guests for beaches, casinos, and other Mediterranean diversions" (Endy, 2004, p. 104). While the Côte d'Azur clientele was democratizing, affluent visitors continued to visit. An American visiting at the time of the 1948 Marshall Plan, which offered American funding to help Europe's post-war economic recovery, complained of "luxury spending and gambling at high stakes in Riviera" by the French and Italians and wondered whether Americans were paying for this. Another reported having seen "more diamonds, furs, luxury yachts and expensive cars than I have ever seen at home." The writer argued that helping the poor in France should be the work of the local wealthy and not the American taxpayer (Endy, 2004, p. 111).

The ethos of *"bouffer, bronzer, et baiser"* was also popularized by the Club Med [Club Méditerranée], established in 1950 by Gérard Blitz, a Belgian diamond cutter and champion swimmer, and his sister Didy. While the Club Med's activities were not limited to southern France -its first vacation villages were in Mallorca- it expanded the imagery of sun-tanning at the beach that was spreading among middle and working classes in the post-war years. By the 1970s, the Club Med's hedonistic *"bouffer, bronzer, et baiser"* imagery had shaped a new image of the Mediterranean in particular and also the seashore in general, so different from the elitist views earlier in the century (Gordon, 2003, p. 15). The summer season had surpassed that of winter with figures for the Alpes-Maritimes at 800,000 visitors in the summer in contrast to 350,000 during the winter by the late 1950s. In 1992, the occupancy rate of Nice's hotels was 41 % in February and 80 % in August (Schor, 2010).

Addressing what they saw as negative images of tourists, a group of authors, associated with the Équipe MIT *(Mobilités, Itinéraires, Territoires)* at the University of Paris 7 – Denis Diderot, linked the sun, sand, and sea of the Côte d'Azur to sex, arguing that to the classic 3 s's (sun, sand, and sea), an "idyllic program" invented in Florida during the interwar years, a fourth "S" for "sex" had been added during the 1960s. Serge Gainsbourg's song "Sex, sea and sun," an extract from the original score for the film *Les Bronzés,* translated as "French Fried Vacation" by Patrice Leconte in 1978, satirizing life at holiday resorts such as Club Med, they argued, substituted sex for sand (Équipe MIT, 2002, pp. 29–30). The bikini, introduced in 1946 had at first been forbidden on the beach at Biarritz and elsewhere but its use by film stars including Brigitte Bardot in 1953 helped change popular attitudes, followed by the appearance of the topless monokini in 1964 at Saint-Tropez. The Équipe MIT authors saw the heightened sexuality in these films as a reversal of centuries of religious teaching. The addition of sex to sun and sand was linking tourism with depravity, giving it a bad name (Équipe MIT, 2002, p. 31). Despite these images of the beach, intensified with the "diversification" of the jet-set and its "rejuvenation" in the 1960s, featuring the appearance of "show-biz" stars, "more or less undressed" at resorts such as Saint-Tropez, a 2001 survey concluded that while half the respondents had a more active sense of sexuality during the summer than at other times, 78 % of those in couples indicated that they would remain faithful to their partners during vacations. "Contrary to received ideas, vacations are more the occasion to reinforce existing ties within the couple, the family, or the group of friends rather than multiply adventures" (Équipe MIT, 2002, p. 34).

The growing popularity of sun and beach was such that in 1978 *Time* magazine reported that ever larger numbers of Europeans traveling on summer vacations to the Mediterranean were overloading local spas and the coastal resorts. Destinations such as Spain's Costa del Sol, France's Côte d'Azur, Italy's Capri, and the Greek islands risked being overwhelmed. *Campeurs sauvages* [unlicensed campers], unable to find accommodations, were pitching tents wherever they could on the Côte d'Azur (Heliomania, 1978, pp. 30–32). By 1985, the Mediterranean coastline attracted 100 million visitors, according to a U. N. report, and had become the world's most popular tourist destination (Urry, 2000, p. 60).

In his *Le Tourisme en France,* first published in 1984 and republished in a fifth edition in 1995, Georges Cazes noted that more than 25 million people, French and international, annually visited the coastal communes, which comprised less than 4 % of France's territory and 10 % of the French population. At the beginning of the 20th century, one French person out of 400 took vacations at the sea. This figure had increased to one of 40 at the time of the legislation of paid vacations in 1936, and one in four at the time of Cazes's writing. Many of the visitors were seeking camping sites but others rented secondary residences, extending the *villégiateur* practice of earlier times. Argelès-sur-mer, near Perpignan on the Mediterranean coast, counted 30,000 campsites or six times the numbers of the local population (Cazes, 1995, pp. 68–69). The "fevered demand" for secondary residences along the coast, in Cazes's words, represented 70 % of total housing capacity in the region. A total of 400,000 "leisure lodgings," which ranged from individual units to vast collective subdivisions or "marinas with feet in the water," was increasing at a rate of 15,000 to 17,000 per year. The number of boats had grown from 20,000 in 1950 to 735,000 in 1990, of which three-quarters were smaller than five and one-half meters in length (Cazes, 1995, pp. 69–70). Stays were also getting shorter with the average reduced to 6.2 nights in 2007 compared with 8.4 nights in 1988 (INSEE, 2008, p. 238).

A report prepared by the Côte d'Azur Comité Régional du Tourism (CRT) in 2018, giving figures for the year 2017, which defined its area as the French Département des Alpes-Maritimes and Monaco, excluding Côte d'Azur – Var, offered figures of 11 million tourists and 70 million overnight stays with an average of 200,000 visitors per day. In words paralleling the concept of the tourism imaginary, an editorial in the report referred to "the region's *brand* [my italics and emphasis], encompassing the Département des Alpes-Maritimes, a part of the Var and the Principality of Monaco" (Côte d'Azur France Tourism, 2018, p. 3). The CRT report's data listed 5.3 million French tourist stays and 5.7 million foreign stays. Foreigners represented 50 % of tourism to the region and 25 % of the tourists were first-time visitors. The year 2017 also saw 6 billion euros spent in tourism consumption, generating 10 billion euros in revenue.

75,000 jobs were created related directly to tourism. Its contribution to the region's economy exceeded 15 %, in contrast to a figure of 7.4 % in France as a whole (Côte d'Azur France Tourism, 2018, p. 5).

The summer season had clearly surpassed the winter, with August the top month, representing 14 % of the total year's stays and a peak on the Assumption Day holiday weekend of 15 August with 650,000 tourists. Continuing a pattern dating back to the 17th and 18th centuries, some 180,000 secondary residences were tabulated for the Alpes-Maritimes Department, of which 47,000 were owned by foreigners. With 15 % of the national total, the Department ranked first in France for the number of secondary residences owned by foreigners. Some half of the tourists were French with one quarter arriving from beyond Europe. The CRT report also listed foreign sources of tourism to the Côte d'Azur in 2016–17 by visits with Italy in first place, the U.K. and Ireland together in second, the United States third, and Germany in fourth position. For stays in hotels and tourist residences, the U.K. and Ireland ranked first, continuing the pattern traced in the early days of international visits to the Côte d'Azur. Presumably, coming from nearby, many Italian visitors either did not stay overnight or did so at the homes of friends and relatives (Côte d'Azur France Tourism, 2018, pp. 6, 9, and 13). Even the Covid-19 pandemic of 2020 did not stop summer vacation travel to the coast. While hotels in Paris were approximately one-third full, the occupancy rate for the Brittany coast was 63 % between 1 and 25 July 2020. In addition, 22 % more French tourists visited the Provence-Alpes-Côte d'Azur region from 11 through 17 July 2020 than during the same period in 2019 (AFP, 2020).

6 Conclusion – Sun, sand, and surf – A tourism imaginary

The continued expansion of Sun, Sand, and Surf tourism along the beaches of France in the 21st century is but the latest chapter in the history of tourism, vacationing, and their imaginaries there. This history, as has been noted, may be traced back to visits to the coastal regions by the wealthy, largely from England and northern France but also from elsewhere in northern Europe, who sought secondary residences in the mid-18th century. The *villégiateurs*, or vacationers, helped give coastal areas such as Dieppe in Normandy and the Côte d'Azur, especially the latter, an allure that was focused on its mild climate, giving rise to its high fashion status during what came to be known during the 19th century as the "season" in winter. France's acquisition of Savoy, following the 1859 Austro-Sardinian War, led to the development of the Nice, Cannes, and Saint-Tropez areas, helping to put the Riviera on the tourist map. Increased railway and later automobile travel, made the coastal areas more accessible for shorter stays and the democratization of tourism, especially with the paid vacations legislation of 1936 added to coastal tourism. Films, such as Maurice Dekobra's "Le Train Bleu" and novels, notably *Tender Is the Night* by F. Scott Fitzgerald, burnished the tourism imaginaries of the Riviera during the interwar years. Of equal, if not greater, importance were shifting attitudes toward skin colour, with pale replaced by a growing fashion of *bronzage,* or sun-tanning, among the upper and middle classes, beginning in the early 20th century. This shift, related to changing perspectives about the medicinal benefits of the sun, as well as evolving cosmetic fashions, both discussed in this essay, is a striking example of the intersections of tourism with other cultural trends in history.

It may be difficult to measure the popularity of an image but the CRT statistics of 2018 show a tourist imagery of Sun, Sand, and Surf in France, as elsewhere, in full expansion during the second half of the 20th century and into the 21st. The report does not break down the activities of the visitors to the Côte d'Azur but its cover is highly significant, showing views of the coast, a skier, and two couples frolicking on a beach. Even an event as harmful to tourism as the Covid-19 pandemic did not weaken the imagery of Sun, Sand, and Surf in France. It may well be said that more than half a century after the May Days of 1968, some of the revolutionaries did, in fact, discover the beach beneath the cobblestones.

References

AFP. (2020). Tourisme. Villes vides, plages bondées ... Un drôle d'été en France. *L'Est Républicain.* https://www.estrepublicain.fr/sante/2020/08/01/villes-vides-plages-bondees-un-drole-d-ete-en-france. 1 August.

Amirou, R. (2012 [1995]). *L'imaginaire touristique.* Paris: CNRS Éditions.

Babulle, R. (1954). *Essai sur le tourisme.* Limoges, France: Rougerie.

Berger, A. A. (2013). *Theorizing Tourism: Analyzing Iconic Destinations.* Walnut Creek, CA: Left Coast Press.

Blanchard, L. R. (1946). *See France.* Rochester, New York: The Gannett Newspapers.

Bonnard. (1927). L. *Le Voyage en France à travers les Siècles.* Paris: Touring-Club de France.

Boyer, M. (1996). *L'invention du tourisme.* Paris: Gallimard.

Boyer, M. (1999). *Histoire du tourisme de masse.* Paris: Presses Universitaires de France.

Boyer, M. (2002). *L'Invention de la Côte d'Azur: L'hiver dans le Midi.* La Tour-d'Aigues: Éditions de l'Aube.

Calasibetta, C. M. (1988). *Fairchild's Dictionary of Fashion*. New York: Fairchild Publications, 2nd edition.

Calmeilles, P. (2018). Mai 68: aux sources d'un slogan mythique: Bernard Cousin a raconté la naissance de son slogan mythique dans un livre paru en 2008. *La Nouvelle République* (16 April 2018). https://www.lanouvellerepublique.fr/indre-et-loire/commune/montresor/aux-sources-d-un-slogan-mythique.

Cannes: The Birth of an International Resort. (1995). *Virtual Riviera*. http://www.riviera.fr/cannhist.htm.

Cazes, G. (1995). *Le tourisme en France* (Series que sais-je?). Fifth edition. Paris: Presses Universitaires de France.

Côte d'Azur France Tourism, Key Figures. (2018). Côte d'Azur, France: Comité Régional du Tourism.

Crowther, B. (1955). Screen: Cat Man Out "To Catch a Thief"; Grant Is Ex-Burglar in Hitchcock Thriller. *New York Times*, 5 August.

de Courcy, A. (2019). *Chanel's Riviera: Life, Love and the Struggle for Survival on the Côte d'Azur, 1930–1944*. London: Weidenfeld and Nicolson.

Deprest, F. (1997). *Enquête sur le tourisme de masse: L'écologie face au territoire*. Paris: Belin.

Ehrenberg, A. (1990). "Le Club Méditerranée 1935–1960." *Autrement. Les Vacances. Un Rêve, un produit, un miroir. Série Mutations*, 111, January, 117–129.

Endy, C. (2004). *Cold War Holidays: American Tourism in France*. Chapel Hill and London: University of North Carolina Press.

Équipe MIT. (2002). *Mappemonde. Tourismes 1: Lieux communes*. Paris: Belin.

Gillespie, N. (1973). All That the Riviera Is, It Owes to Tobias Smollett. *New York Times*, 20 May, Section 10, Travel and Resorts.

Gordon, B. M. (2003). "The Mediterranean as a Tourist Destination, from Classical Antiquity to Club Med." *Mediterranean Studies* 12, 203–226.

Gordon, B. M. (2011). "The Evolving Popularity of Tourist Sites in France: What Can Be Learned from French Statistical Publications?." *Journal of Tourism History*, 3:2, 91–107.

Gordon, B. M. (2018). *War Tourism: Second World War France from Defeat and Occupation to the Creation of Heritage*. Ithaca, NY: Cornell University Press.

Graburn, N. H. H. (1989). "Tourism: The Sacred Journey." In V. L. Smith, ed., *Hosts and Guests: The Anthropology of Tourism*. 2nd edition. Philadelphia: University of Pennsylvania Press.

Heliomania on the Med. (1978). *Time*, 112, 21 August.

Hoerner, J.-M. and Sicart, C. (2003). *La science du tourisme: Précis franco-anglais de tourismologie/The Science of Tourism: An Anglo-French Precis on Tourismology*. Baixas, France: Balzac.

INSEE. (2008). *Le Tourisme en France*. Paris: Institut national de la statistique et des études économiques.

International Tourism Highlights. (2019). *United Nations World Tourism Organization*. https://www.e-unwto.org/doi/pdf/10.18111/9789284421152.

Kanigel, R. (2002). *High Season: How One French Riviera Town Has Seduced Travelers for Two Thousand Years*. New York: Viking.

Karr Alphonse (1808–29 September 1890). (2021). Cimitière Alphonse Karr de Saint-Raphaël (Var). Tombes et Sepultures dans les cimitières et autres lieux. http://www.tombes-sepultures.com/crbst_944.html.

Lanquar, R. (1995). *Tourisme et Environnement en Méditerranée, Enjeux et Prospective, Les Fascicules du Plan Bleu #8*. Paris: Economica.

Levenstein, H. (1998). *Seductive Journey: American Tourists in France from Jefferson to the Jazz Age*. Chicago and London: University of Chicago Press.

Maczak, A. (1995). *Travel in Early Modern Europe*. Cambridge, U.K.: Polity Press.

McKenna, B. (2009). "Melanoma Whitewash: Millions at Risk of Injury or Death Because of Sunscreen Deceptions." In M. Singer and H. A. Baer (Eds.), *Killer Commodities: Public Health and the Corporate Production of Harm* (pp. 145–174). Lanham, MD: AltaMira Press.

Meyer, L. W. (1953). *Driving Through Europe*. San Francisco: Two Books Company.

Nelson, M. (2007). *Queen Victoria and the Discovery of the Riviera*. London and New York: Tauris Parke Paperbacks.

Oxford English Dictionary. Online. (2021). Oxford. Oxford University Press.

Peyroutet, C. (1995). *La France touristique*. Paris: Nathan.

Rauch, A. (1996). *Vacances en France de 1830 à nos jours*. Paris: Hachette.

Salazar, N. B. (2012). "Tourism Imaginaries: A Conceptual Approach." *Annals of Tourism Research*, 39:2, 863–882.

Schor, R. (2010). "History of Tourism on the French Riviera." *NiceRendezVous*, Serre Éditeur. https://www.nicerendezvous.com/history-of-tourism-on-the-french-riviera.html. 6 January.

Shaw, G. and Williams, A. M. (1994). *Critical Issues in Tourism: A Geographical Perspective*. Oxford and Malden, MA: Blackwell.

Smith, V. L. (1989). *Hosts and Guests: The Anthropology of Tourism*. Philadelphia: University of Pennsylvania Press.

Sun-tanning. (1913). *The Times*. 4 September. https://en.wikipedia.org/wiki/Sun_tanning.

Trésor de la Langue Française, Dictionnaire de la langue du XIXe et du XXe siècle, volume 4. (1975). Paris: CNRS.

Trésor de la Langue Française, Dictionnaire de la langue du XIXe et du XXe siècle, volume 9 (1981). Paris: CNRS.

Urbain, J. D. (2003). *At the Beach*, trans by Catherine Porter. Minneapolis and London: University of Minnesota Press.

Urry, J. (2000). *The Tourist Gaze: Leisure and Travel in Contemporary Societies*. London: Sage.

Wilkinson, S. (2012) "A Short History of Tanning." *The Guardian*. 19 February.

Williams, A. (1992). *Republic of Images: A History of French Filmmaking*. Cambridge, Massachusetts: Harvard University Press.

Winter cruising in a summer sea. (1923). *Travel Magazine*, 42, November.

Julie Manfredini

Transformations of tourism. On the French Riviera since the 1950s

Abstract If the *Côte d'Azur* has long been an emblematic destination of French tourism, first in winter, then in summer from 1925, this territory has been marked by mass tourism only since the end of Second World War. The experience of local tourism actors has enabled the establishment of policies for regulation, management and preservation of the environment, especially since the 1960s. By multiplying initiatives in favor of tourism, these actors have become aware of the negative impacts of this activity. They changed their goals for a more environmentally friendly tourism. Thus, if today's saturation of the coast continues, natural resources are now managed to a better use thanks to new initiatives.

1 Introduction

The *Côte d'Azur* acquired its name thanks to Stephen Liégeard and his book published in 1887. Eventually, the coast of Hyerois, the Esterel Massif and the nearest Apennines foothills became the ´gate to Italy´ for early visitors. The Côte d' Azur was described by Liégeard in his book as wild nature embraced by a sea area on the one hand and a mountainous one, on the other, which today causes no terror or repulsion but rather curiosity and fascination. The author came from Dijon and served as deputy prefect at the time. He paraphrased the original name "Cote d'Or" of his home region, forging a new identity for the area which contained part of the Provence and the Nice area along the Mediterranean coast, producing one of the strongest images of the region (Berthier, 2002, p. 236). Thus, "de cette plage baignée de rayons qui mérite notre baptême de Côte d'Azur" (Liégeard, 1887), this definition became one of the first touristic brand names in this territory. Therefore, the success of Liégeard's book is undeniable; upon its re-edition in 1894, Liégeard himself was justified for his choice of name by tourists and inhabitants, undoubtedly thankful to comments written by aristocrats during their "*Grand Tour.*" Thus, *Côte d'Azur* became a geographical designation of this stretch of coastline. Nevertheless, all those who use the expression *Côte d'Azur* do not refer to the same restricted geographical area. From our point of view, *Côte d'Azur* alludes to the part of the coast between Menton and Hyères, which belongs to the French part of the French-Italian Riviera, the real cradle of modern tourism on the Mediterranean coasts and subsequently a large international tourist centre (Callais, 2017b, p. 5). The portion of Genoa is not included in the present chapter.

Until after the Second World War, tourism was concentrated on the *Côte d'Azur* coast during winter, at Hyères or Nice, before the coast became swarmed with holidaymakers (Boyer, 1996, p. 71). Indeed, in the second half of the 19th century, Provence and *Côte d'Azur* were visited by English aristocrats on their *"Grand Tour"* and Russian visitors founded their community/quarter in Nice. This coastal area became a tourist destination without strain although local actors competed to develop comfortable and fashionable activities such as a casino or a racetrack. Even if all coastal resorts did not have the same reputation, they all tried to create walks and opened palaces in order to become attractive for tourists. Indeed, the biggest change occurred on this coast in 1925, even if the local newspapers censored it. The summer season was introduced thanks to American visitors who came to Juan-les-Pins, provoking a cultural shift and the reversal of seasons. This change was, for Marc Boyer, a "mutation contemporaine de l'art d'être touriste," predictable by the evolution of society such as the construction of imaginaries and the adoption of leisure notion by popular categories. So, we are witnessing "l'invention de la Méditerranée estivale" (Boyer, 2005, p. 305). Nice was gradually giving up its envied title of *"capitale d'hiver"* to start, in competition with other large seaside towns, the hunt for this new bourgeois clientele. Tourism became a large-scale activity on this area, especially since the birth of summer tourism.

However, *Côte d'Azur* is in political, economic and cultural competition with Provence, dominated by Marseille. One of the few areas of agreement and collaboration between these two territories remains tourism. Hence, our theory advancing the existence of a tourist territorial coherence from the end of 19th century. Local actors, in particular the local tourist offices, allowed this coherence before the birth of the PACA region. As Marc Boyer points out, this territory is not a historical region, since it does not coincide with Nice County, nor a natural region (Boyer, 2005, pp. 6–7). Rather, it is a "thematic region" which under the pen of Stephen Liégeard appeared as a French Riviera, likely to compete with that of Italy, although its description did not refer to history and neglected the already very marked identity of this territory (Berthier, 2002, p. 439). This was the 19 April 1941 law that formalized the place of the Côte d'Azur in the French landscape and reinforced the initiatives taken so far (Guillon, 2003, p. 426). France is divided into regions, including six for the free zone, headed by prefects. Consequently, some spaces were attached to their traditional territory, according to the Vichy concept and despite their wishes, such as the Var linked to Provence. The regional scale is extremely interesting for tourism development, despite the Vichy government experience, and local tourism actors now claim it. Indeed, in 1947, Jean Monnet's program and the publication of Jean-François Gravier's book entitled *Paris et le désert français* raised awareness on the need for regional economic planning (Duchêne, 1986, p. 72). In August 1955, nineteen program regions were defined, increased to twenty-two in 1956, after negotiations. The Provence-Alpes-Côte d'Azur and Corse region, created in 1956, consequently became the "fruit artificiel de l'arbitraire administratif," because, as Roger Duchêne argued, "l'unité du pays n'a jamais été spontanée" and will only come about with the birth of a region where two rival cities must work together (Duchêne, 1986, p. 47). Tourism provided the connection between the rival cities. From 1 January 1970, the PACA region took the identity Shape that we know today, Corsica breaking away from this territory. The coast of the Côte d'Azur is part of this larger area in the PACA region.

Post World War II tourism reinforced the already strong attraction for the sea on a population enjoying the first paid vacation, which could not really have existed before the Second World War. Tourism actors had to adapt to a summer activity largely practiced by middle-class and popular social strata, less spendthrift and lovers of the outdoors, a situation explaining the success of camping (Gaignebet, 1978, p. 74). In 1947, Nice regained an annual foreign-tourist input equal to that of 1938, numbering around 370,000 people on a narrow strip of the coast. In 1950, visitors were already 665,000 staying in hotels in Nice, Cannes, Menton and Antibes. Tourism flows grew exponentially during the summer season, domestic customers being the largest group: 54 % were French, 9,7 % were British, 9,4 % were Belgian and 8,5 % were North Americans (Boyer, 2007). During this period, the coastal part of Côte d'Azur, in particular the varois (Var) territory was subjected to "indétermination laissant le champ libre à la création de stations ex-nihilo" (Bartoli, 2017, p. 111). However, the vigilance of many actors had "circonscrit les projets et limité leur nombre" (Bartoli, 2017, p. 111).

The above non-outstanding, it might be more interesting to look at this summery *Côte d'Azur* from the 1950s to the present days, a period when the area experienced its greatest change. Let us therefore describe the transformations brought about on the coastline during these "Trente glorieuses" in order to fashion the territory for seaside activity, bathing and mass tourism. While the *Côte d'Azur* was included in the larger tourist territory of the Provence-Alpes-Côte d'Azur (PACA), this area had been affected by tourist pressure due to strong tourist attendance mainly in July and August. The pressure had been strong enough so that the territory had to adapt quickly. Consequently, its evolution was logically connected with its young tourist history made up of multiple identities, allowing it to become the second most important French tourist region in terms of international overnight stays after Paris and the Ile-de-France. Its image had been forged over decades thanks to the name given to the area by Stephen Liégeard but also by many texts and works produced by authors and artists who have praised the advantages of the *Côte d'Azur*, for example, J. Prévert or Nicolas de Staël, making it more attractive. Therefore, the large number of tourists, who are mainly visiting the area in July and August, has never stopped putting pressure on this coastline. This was in contrast with the tourist decrease in the Alpine departments (*Plan de croissance de l'économie touristique*, PACA, 2016). This area was attracting more and more international visitors particularly from new industrialized countries such as Brazil, China and India. The *Côte d'Azur* characterized by tourist pressure and climate variations has been able to adapt over a period of seventy years. With the unquestionable growth of the importance of coasts, the *Côte d'Azur* was particularly turned towards summer tourism, which led to the creation of mass accommodation as the territory was restricted. Despite efforts to balance the flows between the coast and the hinterland, it is argued that the coast remains the main tourist centre. Lastly, as tourist flows balanced on the coastline, new infrastructures have produced a catastrophic artificialization of the coastline.

2 The arrival of mass tourism, an unprecedented upheaval

After the Second World War, tourism resumed on the French Riviera which regained its dynamism around 1947–1948 (Callais, 2017a, p. 9). This territory became easily accessible thanks to urban transport development policies, particularly from the outside world such as the Nice-Vintimille line opened in 1945. Air transportation also opened this area when the Nice airport was created in 1949. The number of passengers rapidly increased, generating an interest for tourism investment in the area. Thus, in 1950, the number of tourists was twice as high as during the period before the Second World War, and welcoming the masses became a common feature of tourist activity. In this context, summer tourism became established and in the 1950s–1960s swimming was a common pleasure for the lower and middle classes. Thus, a tourism season already formed in 1925 on the *Côte d'Azur* continued to flourish while the winter season declined. Infrastructure was adjusted to a growing demand and in some cities hotel industry disappeared before reinventing itself like in the case of Nice which lost 1450 rooms between 1947 and 1957 (Callais, 2017a, p. 12). The coast was the first tourism territory on the PACA region which led to the first tourism concentration plan. The Beach became a place for spending time on the seaside, for resting, having a picnic, getting a tan, reading, playing beach games and getting initiated to swimming. Still, coastal development on the *Côte d'Azur* was unequal because of its important regional historic heritage, and tourist identity. The Alpes-Maritimes were an overexploited coastal strip marked by luxury, while the Var coastline partly retained its original, almost wild aspect. (Casevitz, 1947, pp. 102–111).

Evidence of overcrowding on the coast was caused by fast-growing tourism in the area. For example, it should be noted that 83 % of the Var population in 1954 lived in towns when the urban population of France was estimated only at 55 %. This was in part due to "heliotropism" as many new vacation homes were constructed in areas securing a sea view. This situation pushed up land values and made people envious. New buildings took many forms such as apartment blocks, suburban homes or mobile homes on campsites. Consequently, the housing and automobile boom of the "Trente Glorieuses" years completely upset the balance of the Riviera coast. Development of new hotels and restaurants took precedence over the preservation of nature, in the context of increasing tourism flows (Boriosi, 2017, p. 72). It became necessary to adjust new buildings and their prices to the targeted client groups, particularly the middle classes. This also explained the emergence of a middle-range hotel network (one-star or two-star hotels).

Nevertheless, policies for local development failed to anticipate urban and tourism growth. Thus, the 1940 road system was still being used in the 1950s moreover new construction was considerably delayed, leading to housing shortage problems. It should be remembered that local authorities, as in Nice for example, had remained rather inactive regarding housing and transportation improvement. In contrast, major development like the Nice airport were taking place, raising expectations as Virgile Barrel, a member of the national assembly confirmed: "*La création de l'aérodrome [de Nice] permettra, dans des conditions satisfaisantes, la réalisation d'une entrée digne de Nice par la Promenade des Anglais. Des plantations et jardins pourront être prévus avec une ampleur suffisante pour que les touristes aient, dès leur arrivée sur le territoire de notre ville, l'impression de grandeur et de beauté que Nice doit leur donner [...]*" (Jérôme, 2017, p. 4)

The road network, on the other hand, had not yet been improved. The A8 motorway, for example, was launched in 1976 but was not completed until 2010 (Graff, 2017, p. 96). Nevertheless, some local personalities, conscious of the economic advantage involved, turned to tourism. This was the case of Jean Medecin, influential politician in Nice and tourism expert who focused on embellishing tourist sites, for example renovating the *Promenade des Anglais*. He improved and completed the infrastructure and expanded the international scope through changes in the Nice airport, the management of which had been entrusted in 1956 to the Alpes-Maritimes Chambers of Commerce and Industry (CCI). In 1960, the main runway was lengthened to facilitate the reception of the first Caravels, which put Nice at 1 hour 15 minutes from Paris. This sort of infrastructure enhanced considerably the growth of mass tourism on the French Riviera. Finally, tourism activity was diversified particularly through congress and business tourism. Accordingly, infrastructure was adapted, Nice for example acquiring an exhibition center in 1957, with a reception capacity of around 20 000 persons (Jérôme, 2017, p. 63). Despite these efforts, however, the French Riviera's attraction began to falter, even though G. Jourdan continued to remind:

> Le positionnement métropolitain et la notoriété internationale de la conurbation s'appuient historiquement sur le tourisme. Ce rayonnement touristique international est conforté par l'accueil de nombreux salons, congrès et évènements internationaux ; il explique notamment la forte dimension internationale de l'aéroport de Nice Côte d'Azur. (Jourdan, 2005, p. 11 in Christofle & Hélion, 2018, p. 11)

Thus, in the 1950s, major coastal towns like Nice, Cannes or Menton reaffirmed their importance. *Côte d'Azur*'s attraction was based on their economic positioning, on expertise and recognition of that expertise from tourism actors. After the Second World War the coastline acquired essential assets with which to adapt to tourism and its trends, such as the creation of the first Palais des Festival for the Cannes Film Festival, which in 1947 was acclaimed in international fora. This territory gradually offered a cultural program, which considerably strengthened its reputation and revived its attractiveness, for instance with the international market in music and music publishing (MIDEM), the international market for real estate professionals (MIPIM), the Grand prix de Monaco, the Monaco Circus Festival or the Nice Carnival.

Thanks to the new international airport in Nice, the French Riviera is now almost as easily accessible as Paris. The Nice-Monaco line annually transfers ships, by helicopter nearly ten million travelers, not counting the rail and motorway networks. There is an important hotel park of 732 establishments and 30 000 rooms capacity, 12 palaces, 80 hotels of a 4–star category, or 22 % of the same category nationally (Plasait, 2007, p.17). To this, camping sites like those at Villeneuve-Loubet or Roquebrune-Cap-Martin, should be added, where workers and bosses enjoy spending leisure time (Boriosi, 2017, p. 73). Thus, in 1951, the local press referred to the holidays of a major shareholder of one of the main English shipping companies (*Nice-Matin*, 1951). Furthermore, the president of the Corse-Côte d'Azur Federation of SI confirmed the situation in 1960:

Contrairement à ce que l'on croit, le camping n'est pas seulement un tourisme de bon marché ; il vous suffit de circuler dans les camps de la Côte d'Azur, entre Toulon et Vintimille, pour vous rendre compte qu'il y a des voitures américaines, des caravanes qui valent très cher, simplement parce qu'il y a des gens qui veulent profiter de la mer et rester avec une espèce de petite maison qu'ils emportent avec eux. Il faut que les gens qui veulent faire du camping puissent trouver des conditions qui leur conviennent parfaitement. (Vincent Paschetta in Cavalié, 2013, p. 123)

Undoubtedly, the land-use planning program introduced by the DATAR in 1958 had been successful. This organization had two missions: to organize regional planning and retain foreign tourists passing through the Mediterranean coast. Regarding Hyeres, according to Odile Jacquemin, there have been more transformations: the port was expanded, hotels replaced healthcare institutions, the airport replaced the old naval air base (Jacquemin, 2017, p. 130). Every location on the *Côte d'Azur* has been affected by these changes in order to adapt to tourism flows and new directions given to tourist activity.

3 The new trends in 1960s/70s

With the upturn of economic activity, marketing to promote destinations was accelerated. The *Côte d'Azur* became a dream location for French and foreigners thanks to tourist posters, postcards but especially through the many cinematographic works which staged action there and fixed the image of the destination: with films such as Roger Vadim's, *Et Dieu créa la femme,* Saint-Tropez became the iconic *Côte d'Azur*. Nevertheless, this coastline was no longer only that of luxury, it was also a place for popular strata attracted by the episodes of *The Gendarme de Saint-Tropez,* the first of which came out in 1964. Be it as it may, with this marketing, the *Côte d'Azur* became in 1965 an important tourism destination, second only to Paris.

An intensive urban policy was imposed on part of the *Côte d'Azur* coastline between Saint-Raphaël and Menton (Boriosi, 2017, p. 84). It must be said that the attraction for the coast has grown over the 1960s, creating resorts from scratch like Port Grimaud. In addition, the success of boating led to the proliferation of marinas, such as in Toulon. Thus, the coastline town suffered transformations mentioned in town planning to promote and create a territory specializing in tourist and seaside activities. Local actors wanted to enhance development, which would respect a global approach of improved management for the territory and its activities. The Alpes-Maritimes department engaged in a large-scale urban program through the ZUP (priority areas to be urbanized). But the coastline is narrow and this caused overcrowding and congestion.

Some projects acquired a considerable and unexpected scale, such as the marina Baie des Anges. At the time, financial projects raised interest and Lucien Nouvel became an investor with a pharaonic project including 70 plots of land totaling 26 hectares (Boriosi, 2017, p. 77). After negotiations between many actors, the project gradually emerged. It included 1600 housing units, shops, car parks and a marina and it took over 25 years between 1968 and 1993 to complete. As it were, the *Côte d'Azur* emerged again as the first French tourist destination after Paris.

This may be explained in terms of expansion of new trends such as the rise, in the 1960s, of cultural movements influenced by the Californian counterculture and its representatives: surfing, Rock and drugs. The geography of beaches changed and so did the activities of the actors, hence the emergence of spots for surfers and nudists' beaches. The standard beach faded away for a while, faced with the protest movement of the late 1960s, and a momentary distancing from mass practices. In addition, an increase of second homes. For example, between 1945 and 1968, 50,000 such houses were added to the French Riviera. This added masses of concrete to the coastline, also affected the mountain, and finally the sprawl of hills near Nice. Lastly, recreational boating became more important, the trend no longer being restricted to the rich. Port infrastructure was transformed to adapt to the phenomenon as in Nice. The 1960s indeed justified their misnomer "les années béton" (Bertho-Lavenir, 1999, p. 409).

The 1970s witnessed new transformations related to the explosion of summer tourism but also an overdevelopment of other events. Nice airport continued being an asset, domestic flights being increased and new ones establishing new connections: with Europe (the United Kingdom and Scandinavia in particular), with Africa and with the Middle East (Callais, 2017c, p. 153). Moreover, the airport extension in 1974 illustrated the increase in tourism and the need of major infrastructure. Added to this was the unprecedented real estate fever that transformed the coastline. However, this urbanization was not uniform on the coastline since the Var had been the least urbanized area, having focused on preserving the environment. In contrast, in the Alpes-Maritimes, since 1975 the department was over-urbanized "à l'exception de quelques sections inconstructibles" (Callais, 2017c, p. 143). The Var was a territory with a several small coastal towns and a few thousand inhabitants. The situation would change only when the summer season would begin. But urban development was unequal and 2 million tourists were present, so infrastructure had to be adapted:

> *La création de ports plaisanciers organisés pour accueillir de nombreux bateaux de tonnage médiocre […] la multiplication des magasins alimentaires de libre-service, voire de magasins populaires […], certes, il existe des endroits réservés, où de luxueuses demeures s'efforcent de préserver les parcs et plages qui leur servent d'espace vital […] c'est de plus*

en plus l'avalanche des français moyens qui procure les recettes, qui anime la Côte et en définitive donne le ton. (Livet, 1978, Chp.V)

The coastline of the *Côte d'Azur* has been strongly marked by tourist infrastructure. Hotels, second homes, and camping sites dominated, although rented rooms or flats were the most frequent form accommodation which distorts numbering. Marinas continued to multiply thanks to new legislation, (law of 1965) allowing the government to use part of the maritime public domain (Calais, 2017, p. 148). So, in 1975, there were "63 ports de plaisance dans le Var et 26 dans les Alpes-Maritimes dont certains intégrés à des Marinas" (Calais, 2017, p. 148).

Now, the *Côte d'Azur* is pursuing development of cultural tourism to counterbalance seaside tourism. In this context, the "contrats de destination" created in France concern also the PACA region with three types: "*Arts de vivre en Provence*," "*Voyage dans les Alpes*" and "*Côte d'Azur terre d'événements*" signed in November 2016. With their specific actions, these contracts reaffirm the main activities. On the *Côte d'Azur*, the actors wish to highlight the 6,000 sport and culture events. The ambition is "d'accueillir 500 000 séjours supplémentaires" and achieve the number of 2 million visitors (PACA, 2016, p.18). The "contrat de destination" reinforces the action led for decades by the public and private actors. However, if the infrastructure and housing offer is in relatively good condition on the *Côte d'Azur*, the area itself is ageing and is shrinking, *de facto* limiting large-scale initiatives such as the organization of big congresses. For example, between 1998 and 2005 a reduction in congresses organized in Nice was noted, their number falling from 28 to 16 while at the same time Lyon increased it from 12 to 25, thus joining the top 40 dynamic towns. Henri Ceran, Director of the "Côte d'Azur Convention bureau" clarifies this situation:

> *Sur la Côte d'Azur, on assiste à une stagnation du nombre de touristes d'affaires et à un décrochage de la destination Côte d'Azur par rapport à la montée en puissance de nos concurrents au niveau national et international qui, en termes d'expositions, de salles de congrès, mais aussi de besoins qui correspondent aux attentes des organisateurs, ont des outils plus performants que les nôtres.* (Plasait, 2007, p. 18)

The *Côte d'Azur* is still searching solutions to increase its attractiveness and this means better adaptability facing tourist trends and extending the seasonality of activities and important events. Now, public and private actors agree on at least one point: seaside tourism is always significant, but it is not sufficient for guaranteeing a place of choice within international tourism. Nevertheless, "nature" in the wide sense of the term, in the context of sustainable development can offer new solutions and renew seaside tourism or even tourist offers on the *Côte d'Azur*.

4 Moving towards a more sustainable tourism?

Indeed, the *Côte d'Azur* as well as the entire geographical territory of the PACA region, has many natural areas and an important biodiversity. It makes sense, therefore, to look for a key environmental concern in local and national tourist policies. The most emblematic measure had to do with the Port-Cros national Park. The Alpes-Maritimes, as well as the entire coastline of the French Riviera, benefits from tourism especially when compared to the Var. However, tourism accentuates difficulties related to water supply (especially during the summer), when coastal areas get crowded contributing to serious pollution and other problems such as erosion. These difficulties become a real handicap when tourist flows increase in the summer season. Some initiatives have therefore been taken by the Canal of Provence Company to improve access to clean water, on the one hand, and on the other, to build the new highways of the Sun and the Provençale without, however, completely solving the problem of traffic jams.

In the 1960s, several projects for the preservation of the coastline were developed, such as the Vaugrenier department park in the Alpes-Maritimes. Some investors reclaimed the seafront offering better protection via a pedestrian promenade created thanks to the decision taken by the Villeneuve-Loubet mayor, Richard Camou (Boriosi, 2017, p. 84). Thereafter, from 1977 onwards, some cities developed pedestrian areas in the city-centre. For example, Nice refused an uncontrolled urbanization and is still keeping the "longue tradition de régulation urbanistique" (Graff, 2017, p. 98). As for the Var area, which belongs to the *Côte d'Azur* appellation, it is particularly active in preserving nature. Hyères, for example, which is one of the best-known tourist destinations on the *Côte d'Azur* along with Nice and Saint-Raphael, is a seaside town where tourism has taken root without, however, altering the wild look of the landscape. Yet, just like its neighbours, Hyères has known the real estate fever in the 1960s but was able to rebound through a policy of environmental and coastal preservation. It should be noted that it had an important asset, that is knowing how to combine 24,000 ha of landscape with 24,000 ha of marine(sea)scape. Thus, despite urbanization projects, such as the Prost development plan of the Var coastline, as early as 1923, a prefect aware of the richness of the territory sought to reconcile conservation with development (Jacquemin, 2017, p. 126). Fortunately, the *Désirade* project of building a lacustrian town inspired by the architect of Port-Grimaud (F. Spoerry) did not materialize thanks to the acquisition of the land by the coastal conservation authority (Jacquemin, 2017, p. 132). The projects planned and implemented incorporate housing in the natural environment, as the VVF la Badine on the Giens peninsula, in the heart of a pine grove, showed.

Facing the city, the Port-Cros national park attempts to preserve the terrestrial and lacustrian environment with 1,700 ha of land surface and 2,900 ha of marine space under its protection. The success of the Port-Cros national park built in 1963 and its preservation is the perfect illustration of the vocation by Hyères city. In 1971, a coastal development plan on the coast of Provence-Côte d'Azur created by the regional equipment department of demonstrated the excesses caused by the wait-and-see attitude of local authorities in matters of construction. It also revealed the future danger of saturation predicted for the following decade, warning signs of which were already visible.

In addition to problems created by tourism, many others persisted such as atmospheric pollution caused by the greenhouse gas emission effect and noise pollution on a coastline overloaded by the use of private cars and numerous air flows. With the development of air conditioning, tourist residences have become large consumers of energy. The negative effects of this urbanization increase are getting worse and the risk of climate hazard is exacerbated. The Alpes-Maritimes came close to it in 2015 and in 2020 when floods in the area were blamed on concretization. Obviously, the climate risk is exacerbated by this situation. Mass tourism also increases the production of domestic waste which must be recycled. Deterioration of water quality is also noted in coastal sea waters in the summer season. Ways to solve the problem were tried since the 1960s, with hydraulic development works carried out by EDF and the Canal of Provence Society. They involved using the water of the Durance and Verdon, on the one hand, and on the other, to create reservoirs such as the Serre-Ponçon dam in 1961 or the Castillon, Sainte-Croix et Gréoux reservoirs (Callais, 2017c, p. 150). Since 1975, the Canal de Provence supplies the coastline of Toulon and the French Riviera on the side of the Var with water, while the west of the Côte d'Azur is supplied by the Saint-Cassien dam and underground waters for the Grimaud Saint-Tropez sector (Callais, 2017c, p. 150).

A strong artificialization of the coast has been observed due largely to "heliotropism" and the maritime attraction of the coast. The construction of houses as well as port infrastructure geared towards the search of sea views saturated the coastline. As Samuel Robert notes, artificialization develops three times faster in an area with a view (Robert, 2009, p. 296). Thus, we note an over frequentation of these natural sites, especially near large cities, several places exceeding the 800,000 visitors per year. Trampling, uncontrolled extractions, destruction of Posidonia seagrass by private pleasure boats mooring in a bay are some of a series of aggressive acts against nature. Furthermore, overcrowding on the Côte d'Azur coast worsened two major problems which local actors should remedy: beach and sea pollution in areas frequented by tourists, as well as the disappearance of agricultural land because of the real estate fever (Samak, 2016). As Alain Callais recalls, sewers are insufficient given the size of the flow of people and household waste is growing exponentially. Boaters have caused such marine pollution that the closure of some beaches such as Hyères in 1972 became necessary (Callais, 2017c, p. 149). Towards the Alpes-Maritimes it is the CIPALM (Pollution intervention unit in the Alpes-Maritimes) which monitors levels over both air and land. Concerning the agricultural lands, which have been the wealth of our territories since the XVIIIe century, they are gradually disappearing because tourists want to be near the sea. However, these agricultural lands have the advantage of being affordable and already serviceable (Callais, 2017c, p.149). Some areas still resist, such as the municipality of Bormes-les-Mimosas thanks to its famous mimosa festival.

However, since 1970 new laws have been passed to make governments, inhabitants and some tourists aware of the environmental issues. The Chirac circular of 4 August 1976 imposed the maintenance of ecological balances and the limitation of consumption in coastal areas. Similarly, the *Conservatoire de l'espace littoral et des rivages lacustres* in 1975 and the Ornano directive in August 1979 raised an increasing awareness. This directive, for example, has introduced in the Town Planning code the first coastal-specific provisions. Consequently, well before the coastline law of 1986, an awareness conscience is progressively built up. Nevertheless, it was the SRU law (Solidarity and Urban Renewal Act) which in 2000 led to SCOT (Territorial coherence program) and encouraged local actors to mobilize for the future organization of a large geographical area with several cities on the coastline in order to build a global territorial conservation program. Simultaneously, other initiatives proliferated in the hinterland, with similar objectives such as reducing road congestion and promoting "natural" tourism. Tourists could access the Mercantour National Park and the Centre Alpha in Saint-Martin Vésubie. Nevertheless, studies show that in the PACA region, tourist flows continued to concentrate on the coastline while the hinterland received only a small portion of visitors. After all, the sea was the main asset of the area, which was also important for tourists; judging by the actions of the Métropole Nice Côte d'Azur, local authorities wished to protect this territory (Métropole Nice Côte d'Azur, 2021). For example, the first objective of this actor was a better control of sanitation and discharges at sea, hence the appearance of the ports of azure, eight marinas creating a network respecting an environmental charter (see Illustration n°1). They were part of the "blue policy" of the Riviera, an innovative policy that took into account territorial coherence, improved quality of life on a "sustainable" coast and made the port a "porte d'entrée maritime pour la métropole." Similarly, the Natura 2000 "Cap Ferrat" site, which belongs to European Natura 2000 network and is 100 % marine, was created in 2009 in accordance with the European directive « Habitats, Faune, Flore » covering 9000 ha (see Illustration n°2). This territory lies between Cap Ferrat and Cap d'Ail and its objective is to preserve the exceptional marine ecosystem that exists in this area. Henceforth, the Posidonia meadow, the Mediterranean marine plant that cannot be found anywhere else, and the bottlenose dolphin belong to the species that this program intends to protect. For its part, the Port-Cros National Park has been arranged according to the new directives. Since 2012, the Park comprises two "hearts" that is to say parts including the island of Port-Cros, areas owned by the State and a marine fringe over 600m. In addition, a sustainable development policy is carried out in conjunction with local municipalities in an "area of membership" (La Garde, le Pradet, Hyères, La Croix-Valmer and Ramatuelle). Finally, an "adjacent maritime area" reproduced at sea in the accession aid, is extended over three nautical miles (Port-Cros national park, nd). The park works indirectly to attract tourists by enhancing the natural and cultural landscape heritage of the places, such as the restoration of the Fort du Moulin in Port-Cros initiated in 2012 (Port-Cros national park, 2018, p. 8). There are multiple approaches both in form and location since even the Levant and Giens may be affected by initiatives such as the landscape project to enhance the gardens of the Levant or the redevelopment of the Pradeau Fort in Giens. The dynamism of the place makes it possible to perpetuate this structure. Furthermore, the cultural program operating for several years reinforced the charm, with the "Jazz aperitif" during the "Jazz à Porquerolles" festival in 2018 where we had the opportunity to show the capacity and the evolution of tourism in Porquerolles (Port-Cros National Park, 2018, p. 10). When the Park is properly run, its sustainability is guaranteed, as well as its contribution to sustainable development. In fact, the brand *"esprit parc national Port-Cros"* was created with this in mind. Its development has included opening up to local craftsmen and producers, as well as allowing to advertise winemakers eligible for the brand in 2018 (Port-Cros National Park, 2018, p. 26).

Illustration n°1 : The Ports Azur

Service de la gestion des activités portuaires et maritimes. Ports Azur, the port network of the Nice-Côte d'Azur metropolis. March, 2017.

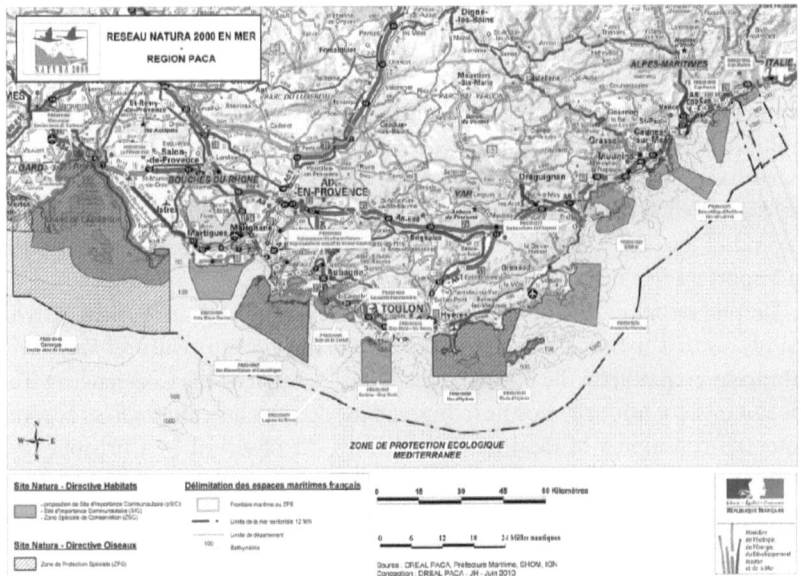

Illustration n°2 : The Natura 2000 areas on the Côte d'Azur

DREAL, Préfecture Maritime, June 2010.

5 Conclusion

The territory, known as French Riviera, a region marked by the development of tourism, mainly for the summer, is almost exclusively oriented towards the tourist economy. Endowed with historical and patrimonial assets and a multifaceted tourist offer (œnotourism, professional activity, health tourism), the *Côte d'Azur* has incited tourist activity by planning coastal areas. Among the main activities, the sea retains a predominant role in the recreational activities, but other opportunities have been integrated such as meetings, congresses or dynamic events, hence the designation of the *Côte d'Azur* as "land of festival." Despite some imbalances and the negative impact of mass tourism, which saturates the coastline, the *Côte d'Azur* is an urban area better developed than other parts of France. Nevertheless, geographical imbalances are noted as well as urban growth. On the all-French Riviera territory, the Var coastline keeps its particularity, as an area where nature is particularly protected and tourist activity developed around this theme. Excessive speculation, however, continues, to the expense of indigenous residences, as out of 170,000 second homes identified in 2012, one third belonged to foreigners. Like any territory geared towards tourism, in the azurean coastline, leisure and tourism activities are prolonged extending the season and generating an almost permanent tourist flow. The area is marked by the diversification of tourist activities and an economy where services predominate. Despite multiple initiatives since the 1970s to restore the balance, by fostering domestic tourism development, in the hinterland, tourism is still largely concentrated on the coast. Heliotropism as a phenomenon that has branded the *Côte d'Azur* since the end of the XIXe century, was further strengthened over time, and still plays an important role in its fame.

Despite recent efforts to unclog the coastline and rebalance tourism activity mainly centred on the coast and seaside activities, the pressure will probably never disappear even despite the unprecedented crisis caused by the Coronavirus. The inhabitants, traditionally hostile to tourism for decades have themselves become potential consumers of local leisure activities. As observed by Christofle and Hélion: "*Le résident est […] de plus en plus intéressé par son cadre de vie et participe ainsi de façon croissante aux événements et festivités, notamment pendant la saison touristique et les week-ends*" (Christofle & Hélion, 2018, p.11).

Recently, this territory has been upgraded in the market " liée à un reclassement des hôtels 5 étoiles" and "un affaiblissement dangereux de l'offre hôtelière économique de qualité" (PACA, 2016, p.14). Investment increased particularly in the case of demand for campsites and for mobile homes. The territory is accredited as top "tourism quality" by the appropriate institutions (PACA, 2016, p. 15). The *Côte d'Azur* has all the assets to have a good position during the phase of tourism recovery. It always had a strong branding, its image reaffirmed by the recent creation of the brand *Côte d'Azur* in 2009. Officially registered with the INPI, this new asset becomes a solid element used by the CRT, which associated it with its activities. This is why the tourism sector is optimistic despite numerous closings and declining job offers compared to previous years. In fact, the sector is already launching seasonal recruitment campaigns for the summer of 2021 (Violaine III, 2021). Thanks to the e-fairs of tourism, it may be possible to consider a slow but certain recovery.

Sources

Casevitz J. (1947). « Le recensement du 10 mars 1946 », *L'information géographique*. vol. 11, n° 3, 102–111.

Archives départementales des Alpes-Maritimes, *Nice-Matin*, 28 août 1951.

Liégeard S. (1887). *La Côte d'Azur*. Paris. Maison Quantin. 443p.

Livet R. (1978). *Atlas et géographie de Provence Côte d'Azur et Corse.* Paris. Flammarion. 291 p.

Violaine III. (2021). « Coronavirus ; malgré la crise, la Côte d'Azur recrute pour le tourisme cet été », *France bleu azur.* [Online]. https://www.francebleu.fr/infos/economie-social/coronavirus-malgre-la-crise-la-cote-d-azur-recrute-pour-la-saison-touristique-estivale-1615324487. Accessed April 2021.

PACA. (2016). *Plan de croissance de l'économie touristique,* schéma régional de développement touristique 2017–2022, 56p.

Métropole Nice Côte d'Azur. (2021). [Online]. https://www.nicecotedazur.org/environnement/mer. Accessed April 2021.

Port-Cros, parc national. « Un territoire reconnu ». [Online]. http://www.portcros-parcnational.fr/fr/le-parc-national-de-port-cros/un-territoire-reconnu. Accessed April 2021.

Port-Cros, parc national. (2018). « Compte-rendu d'activité ». [Online]. file:///C:/Users/jmanf/Downloads/cra2018_lg.pdf. Accessed April 2021.

Samak, M. (2016), « Les Alpes-Maritimes sous pression urbaine, retour sur quarante ans de déclin des surfaces agricoles », *Métropolitiques,* [Online]. https://metropolitiques.eu/Les-Alpes-Maritimes-sous-pression.html. Accessed April 2021.

References

Bartoli, P. (2017). "Les programmes résidentiels de vacances dans la période des Trente Glorieuses : expérimentations et innovations sur le littoral varois." *Recherches régionales Alpes-Maritimes et contrées limitrophes* (212), 109–122.

Berthier, G. (2002). *Dictionnaire de la Provence et de la Côte d'Azur.* Paris: Larousse.

Bertho-Lavenir, C. (1999). *La roue et le stylo, comment nous sommes devenus touristes.* Paris: Odile Jacob.

Boriosi, M. (2017). "Quand le tourisme transforme le paysage : l'exemple de Villeneuve-Loubet pendant les Trente glorieuses." *Recherches régionales Alpes-Maritimes et contrées limitrophes* (212), 69–86.

Boyer, M. (1996). *L'Invention du tourisme.* Paris: Gallimard.

Boyer, M. (2005). *Histoire générale du tourisme, du XVIe au XXIe siècle.* Paris: L'Harmattan.

Boyer, M. (2007). "Introduction au colloque Histoire du travail dans l'hôtellerie et la restauration sur la Côte d'Azur au XXe siècle." *Recherches régionales* (189), 6–7. [Online].https://www.cg06.fr/documents/Import/decouvrir-les-am/recherchesregionales189.pdf. Accessed September 2013.

Callais, A. (2017a). "La Côte d'Azur de l'Après-guerre." *Recherches régionales Alpes-Maritimes et contrées limitrophes* (212), 9–16.

Callais, A. (2017b). "La Côte d'Azur des Trente glorieuses (1945–1975), Introduction." *Recherches régionales Alpes-Maritimes et contrées limitrophes* (212), 3–8.

Callais, A. (2017c). "Qu'est devenue la Côte d'Azur au milieu des années 1970 ?." *Recherches régionales Alpes-Maritimes et contrées limitrophes* (212), 141–158.

Cavalié, H. (2013). *Trois siècles de tourisme dans les Alpes-Maritimes*. [Exposition des Archives départementales des Alpes-Maritimes, décembre 2013–30 mai 2014]. Milan: Silvana Editoriale.

Christofle, S., Hélion, C. (2018). "Territoire(s) et tourisme(s): hybridation des pratiques et des espaces à Nice Côte d'Azur " Métropole" ?." In D. Crozat et D. Alves. *Le touriste et l'habitant* (pp. 1–21). Cergy : Connaissances et Savoirs, Patrimoine et Tourisme.

Duchêne, R. (1986). *Naissance d'une région (1945-1985)*. Paris : Fayard.

Gaignebet, J-B, Guiral, P. (dir.). (1978). *La Provence de 1900 à nos jours*. Toulouse: Privat.

Graff, Ph. (2017). "Un regard urbanistique sur les mutations du tourisme à Nice lors des Trente Glorieuses." *Recherches régionales Alpes-Maritimes et contrées limitrophes* (212), 87–104.

Guillon, J-M. (2003). "L'affirmation régionale en pays d'Oc des années 1940." *Ethnologie française* (33/3), 425–433.

Jacquemin, O. (2017). "Hyères dans la Côte d'Azur des Trente Glorieuses : l'inversion du regard, la fin des marinas et la naissance de la protection de l'environnement." *Recherches régionales Alpes-Maritimes et contrées limitrophes* (212), 123–134.

Jérôme, Ph. (2017). "Virgile Barel et Jean Medecin, deux conceptions de la politique du tourisme." *Recherches régionales Alpes-Maritimes et contrées limitrophes* (212), 59–64.

Plasait, B. (2007). *Le tourisme d'affaires un atout majeur pour l'économie touristique*, Rapport du conseil économique et social. [Online]. https://www.vie-publique.fr/sites/default/files/rapport/pdf/074000449.pdf. Accessed April 2021.

Robert, S. (2009). *La vue sur mer et l'urbanisation du littoral. Approche géographique et cartographique sur la Côte d'Azur et la Riviera du Ponant*. Thèse de géographie. Université Nice Sophia Antipolis.

Carlos Larrinaga

Spain after the Civil War (1936–1939). The New Possibilities for Maritime and Coastal Tourism[1]

Abstract During the first third of the 20th century, an incipient tourist system took place in Spain. The Civil War (1936 until 1939) had a huge impact on that system but it persisted nevertheless. Therefore, during the first Franco regime (prior to 1959), Spain managed to rebuild its tourist system and, thanks to its Mediterranean climate, maritime tourism gained increasing importance in the country's tourism offer. Thus, new tourist areas along the coastlines were added to the traditional tourist areas (San Sebastian and Santander).

1 Introduction. Foundations before the Civil War

During the first third of the 20th century, an early tourist system was being formed in Spain made up of destinations and products, a market, certain agents (tourists or consumers, active and receptive tourism associations, Administration and private companies) and a tourist organization (Vallejo, 2018, p. 85; Vallejo, 2019b, p. 177). Therefore, although Spain lagged behind others in the international tourism landscape, especially when compared to nearby countries such as France or Italy, the truth is that until the outbreak of the Civil War in 1936, the foundations had been laid of what we now consider modern tourism, i.e. what was understood as an industry (Norval, 1936). More specifically, an entire tourism industry was set up around certain businesses such as the hotel industry, restaurants, travel agencies, transport companies or some urban developers which played an important role in creating tourist areas or accommodation, for example.

In those decades, Spain could not yet be considered a touristic country, meaning by this, a country which receives a large number of foreign tourists and whose tourist balance has a surplus. Rather, it can be classified as a tourist country since there is a considerable flow of national tourists abroad and its tourist balance is loss-making. However, Spain can be described as a country

---

1 This study is part of the research project HAR2017-82679-C2-1-P, financed by the Ministry of Science and Innovation of the Government of Spain and the ERDF.

for tourism, as tourism as a social practice and economic activity were gradually gaining weight to such an extent that the Civil War had a negative impact on that afore-mentioned budding tourist system but did not manage to end it. Indeed, the reconstruction of the Spanish tourism system began even in these war years, at least as far as the Franco regime is concerned. All this despite the post-war period and the international context itself (World War II and years before the Marshall Plan), which made this intent very difficult.

Of all the components which make up the tourist system, in this article I will focus on the possibilities of maritime and coastal tourism, a modality that made its appearance on the Atlantic coast of northern Spain in the 19th century and which, over time, achieved a growing diversification, by gradually incorporating new coastal regions, especially in the Mediterranean sea (Barke, Towner & Newton (Eds.), 1996). Actually, due to the many realities or facets that tourism presents, it can be noted that, during the first third of the 20th century, different tourist geographies were formed within the country. This means, we are discussing areas that were progressively specializing in this activity (Gil de Arriba, 2018, p. 173). Coastal areas were the most requested although not the only ones. First, the areas in question are linked to health issues, related to the healing properties of the waters and the benefits of sea breezes. Then, once pharmacopoeia was gaining weight, there were areas for recreational and leisure purposes. And finally, due to that new cult of sun, body and water sports that, at least, lavished in the '20s and strengthened after World War II (Boyer, 2002). Spain, with almost 8,000 linear kilometres of coastline, went through this whole process, although not in all its coastal areas at the same time.

By shaping the different tourist destinations that emerged in the first third of the 20th century, as well as in the routes or itineraries that were being offered, the increasing spatial mobility played a fundamental role (Lickorish & Kershaw, 1958). Whereas railways had been the means of transport *par excellence* in the 19th century, motor vehicles (cars and coaches) were gaining prominence, especially from the 1920s onwards in the next century. On the other hand, in a country with so much coastline, numerous ports and two archipelagos, we must not forget the prominence of ships for transporting tourists, as well as cruising tourism itself. Also, and as a precedent of its subsequent importance, we must mention planes which would gain prominence as the 20th century progressed.

Already focusing on these coastal areas, the region of Cantabria, on the Atlantic coast of Spain, played a decisive role during the reign of Alfonso XIII (1902–1931). In fact, the new monarch merely followed a tradition that came from the mid-19th century, as cities such as San Sebastian or Santander had already been favoured by the presence of the royal family. Alfonso XIII chose both capitals as summer resorts, thus dragging there numerous courtiers, politicians, diplomats and artists, to name but a few. From the last quarter of the 19th century, this model of tourism of cold beaches, characteristic of the entire European Atlantic façade, was developed in this area. Its proximity to the south of France, especially in the case of San Sebastian, also allowed a greater presence of foreign tourists, reaching relatively international fame. In this sense, the proximity of Biarritz proved to be decisive. A coastal area *par excellence*, considered mainly by the upper layers of Spanish society, the Cantabrian region was in direct competition with new tourist areas from the twenties onwards. In a context of transit from cold beaches to warmish beaches, there was, at the same time, the popularization of wave tourism and the introduction of middle classes to tourism. Also, that tone of exclusivity that this coastal area had had was yielding in the twenties, although it continued to maintain a social elegance which other destinations did not achieve (Gil de Arriba & Larrinaga, 2021).

Although the administrative identification of tourist areas by specific names did not occur until the sixties and seventies, already with the reality of mass tourism in Spain, actually in the first third of the 20th century, denominations were coined that little by little made a fortune. Above all, to designate coastal areas of the Mediterranean. Following the appellative model such as the *Côte d'Azur* (French Riviera) or the *Riviera italiana*, the Costa Brava also emerged in Spain on the Catalan coast or the Costa del Sol, at first not as well defined as in the French and Italian case, although they were after the Civil War (Gil de Arriba, 2018, pp. 174–183). These types of destinations and the Balearic Islands, especially Mallorca, began to compete with the classic destinations of the Cantabrian region, tacking that new, afore-mentioned reality. An entire process of *southernisation* of tourism had begun (Larrinaga, 2015), where the prominence of the Mediterranean was evident (Pemble, 1987; Soane, 1993). The enhancement of the Mediterranean coast was creating a new tourist identity, while laying the foundations of those which would become, already during Franco's regime, the main tourist destinations in Spain (Gil de Arriba, 2018, pp. 182–183). In turn, the Balearic Islands, with Palma de Mallorca at the helm, had become an important tourist destination in the western Mediterranean since the late twenties and even more so in the thirties. The different indicators that we have point in that direction (Vallejo, 2018, pp. 140–152), to the extent that tourism became a fundamental activity in the economy of the island before the Civil War (Buades, 2004; Cirer, 2009).

2 Spain as a sun and beach tourist destination

The *southernisation* of tourism or the conquest of the Mediterranean was not exclusive to the second half of the 20th century. In fact, it was a process that had occurred before, especially after the Great War, when the warm waters of southern Europe were more suitable for certain water sports and when the fashion of sun tanning began to increase (Boyer, 2002, p. 28). However, it is true that it was not until after World War II that this phenomenon spread and eventually consolidated, enhancing the beaches of the Mediterranean and causing a very considerable increase in tourist flow to these destinations. The key to such a flow was the increase in the economic capacity of European and North American households, which had the money and time needed to participate in tourist activities. There were several factors which contributed to this, namely: the economic prosperity of North America and a large part of Europe, the increase in personal income of increasingly large sectors of the population, the improvement of means of transport (especially aeroplanes), the general spread of paid holidays and the increase in free time (Pellejero, 2005, p. 92).

By 1950, for example, having already become one of the main industries in Europe, tourism grew proportionately faster in Spain than in any other country. Any flaws presented before then due to transport or the offer of tourist accommodation could be resolved overtime. However, Spain had competitive advantages in terms of weather (sunshine, basically), prices (cheap) and the variety of its attractions (Pack, 2006). Moreover, some travel guides pointed out that Spain was a country that should be visited (Barke & Towner, 1996, p. 17), for example, the Fodor guide (1952). Thus, receptive tourism in Spain grew in the 1950s and 1960s at annual rates of 25.2 and 17.2 %, respectively, doubling the European and world averages. In this way, the beaches of the Mediterranean coast and some islands became the international epicentre of massive coastal tourism, that is, sunshine and beach (Vallejo, 2013, p. 428). (See Tab. 1.)

Tab. 1: International tourist arrivals (annual growth rate %)

	World	Africa	The Americas	Asia and the Pacific	Europe	Middle East	Spain
1950–1960	10.6	3.7	8.4	14.1	11.6	12.3	25.2
1960–1970	9.1	12.4	9.7	21.6	8.4	11.5	17.2
1970–1975	6.0	14.4	3.4	10.5	6.4	13.0	5.2

Source: Vallejo, 2013, p. 428.

Tab. 2: Visitors in southern Europe (in percentages)*

Years	Spain	France	Italy	Others
1950	11.3	45.9	40.0	2.8
1951	13.5	43.2	40.1	3.2
1952	14.9	39.4	41.6	4.1
1953	15.8	34.5	43.8	5.9
1954	15.7	33.8	43.9	6.6
1955	16.9	31.5	44.2	7.4
1956	16.8	31.6	45.0	6.6
1957	17.3	27.2	48.7	6.8
1958	19.7	25.5	47.2	7.6
1959	19.6	27.7	44.3	8.4
1960	24.7	26.8	40.6	7.9
1964	36.0	20.7	31.5	11.8

* Does not include hikers or day-trippers.
Others: Portugal, Yugoslavia, Greece and Turkey.
Source: Fernández Fúster, 1991b, p. 597.

As can be seen in Tab. 2, Spain has been gaining in position since the mid-1950s. In fact, until 1958–1959, the rise in its market share was rather slow. It was in 1960 when the take-off really took place, reaching almost 25 % of the foreign tourists who visited southern Europe (see Graph 1). The so-called miracle of Spanish tourism was therefore reinforced, going from about half a million visitors in 1950 to 10.5 million in 1964, with a market share that already accounted for 36 %, beating both France and Italy that year, which, since the early fifties had been the favourite of the southern countries of Europe (Fernández Fúster, 1991b, p. 598). Moreover, its global positioning was also spectacular: whilst in 1950 it only received 1.8 % of total tourists and 0.8 % of international tourism revenue, ten years later the figures had reached 6.2 and 4.3 %, respectively (Vallejo, 2013, p. 426). A similar influx of travellers had their response in the hotel industry itself. Thus, the number of establishments (hotels and hostels) increased from 1,318 in 1951 to 2,274 in 1958. As for available rooms, it rose from 48,226 to 78,497 during that period (Banco de Vizcaya, 1960, p. 15).

From the political point of view, we must mention, first of all, a certain opening of the Franco regime, which reduced the entry requirements of foreigners (Pack, 2006). And, secondly, that Spain was abandoning the political isolation to which it had been subjected at the end of World War II, something that was further accentuated after the renewal of the government in 1951. Back in 1950, the UN

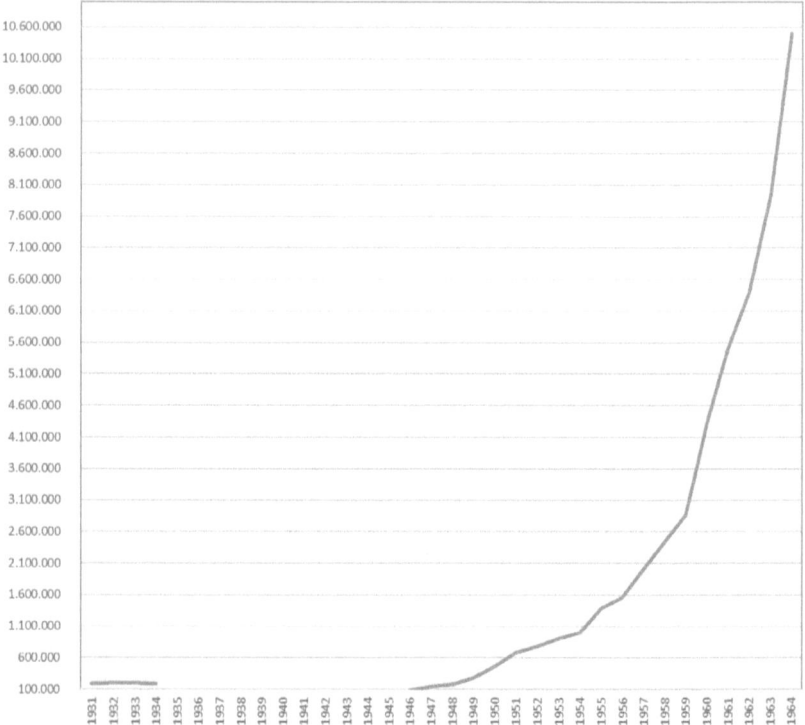

Graph 1: Entry of travellers in Spain, 1931–1964
Note: these are foreigners with passports.
Source: Fernández Fúster, 1991b, Madrid, p. 600.

removed the sanctions that it had imposed on Spain in 1946 and allowed the return of the ambassadors. In addition, between 1949 and 1950, the US government approved the first financial aid items to Spain through three financial entities, namely: the Chase Manhattan Bank, the National City Bank and the Export-Import Bank. To be more specific, in this opening framework, in January 1950 the Director General of Tourism made a trip to the United States and Cuba for two months to promote tourism in Spain, with inauguration in the Spanish Tourism Office of New York, Chicago, Los Angeles and San Francisco (Correyero & Cal, 2008, p. 429). However, this process would culminate with two especially important events: the signing of the concordat with the Holy See in 1953 and the Madrid Pacts of the same year with the United States, under which four

American military bases would be installed on Spanish soil in exchange for economic and military aid. In 1955, Spain entered as a full member of the UN and from that year onwards, it was integrated into the various international financial organizations.

From an economic point of view, in addition to the propaganda that was carried out from the General Directorate of Tourism, the key to this rise must be seen in prices. The Spanish tourism authorities resorted to the containment of prices to gain competition from other countries. Under a dictatorial regime and with an intervened economy, government tourism officials imposed very low prices to make Spain a cheap market. With sunshine and beach guaranteed in the Balearic Islands, in the Mediterranean and in the southern peninsula, they saw the containment of prices as the best way to compete with other destinations that offered similar products, such as Italy or Mediterranean France (Arrillaga, 1955a, p. 159). (See Tabs. 5 & 6). Thus, on 8 August 1946 the Council of Ministers agreed on a preferential exchange rate for tourists: 16.4 pesetas per dollar, while the official price of the peseta in dollars was 10.95 pesetas per dollar between 1941 and 1959. In December 1948, a new special tourist exchange rate of 25 pesetas per dollar was established, something that had been requested by the different tourism agents. However, the most spectacular devaluation occurred in 1959, with 60 pesetas per dollar when in 1957 it had been set at 42. This fact had an immediate impact on the arrival of tourists and, consequently, on GDP (Fernández Navarrete, 2005; Pack, 2006; Vallejo, 2013). (See Tab. 3.)

Tab. 3: Foreign tourism, Spanish population and share in GDP, 1901–1970

	Spanish population (thousands)	Foreign tourists (thousands)	% Tourists/ Population	% Income overseas Tourism / GDP
1901	18,659,0	116.5	0.6	0.7
1930	23,445,0	277.9	1.2	0.4[1]
1940	25,757,0	18.9	0.1	0.1
1950	27,868,0	457.0	1.6	0.4
1957	29,548,0	2,018.7	6.8	0.8
1960	30,303,0	4,332.4	14.3	3.0
1970	33,876,5	21,267.0	62.8	5.1

(1) The share of foreign tourism in GDP corresponds to 1931.
Source: Vallejo, 2013, p. 428.

Tab. 4: Tourist income and foreign trade (millions of dollars)

Years	(1) IEME income	(2) Income according to the OEEC	(3) Value of exports	(4) Value of imports	% which represents (2) out of (3)	% which represents (2) out of (4)
1953	94.2	94.2	596	482	15.8	19.5
1954	90.0	104.8	614	464	17.1	22.6
1955	96.7	143.7	617	446	23.3	32.2
1956	94.8	163.3	767	442	21.3	36.9
1957	76.9	213.4	862	476	24.7	44.8
1958	71.6	264.9	849	486	31.2	54.5

IEME: *Instituto Español de Moneda Extranjera*, Spanish Institute of Foreign Currency.
Note: the differences in figures in columns 1 and 2 are due to the provision of pesetas outside the IEME.
Source: Banco de Vizcaya (1960). "El turismo en España," *Revista financiera*, 85, p. 17.

Tab. 5: Prices of hotels in the different countries in 1953 (by categories)

Countries	In dollars and base 100 = Unites States				
	Luxury	1st A	1st B	2nd	3rd
Argentina	6.09 (32)	4.18 (29)	3.04 (33)	2.66 (48)	1.74 (99)
USA	19.00 (100)	14.00 (100)	9.00 (100)	5.50 (100)	1.75 (100)
France	10.00 (52)	8.28 (59)	4.28 (47)	3.75 (70)	2.00 (114)
England	10.20 (53)	4.47 (31)	3.77 (41)	2.79 (50)	2.09 (119)
Italy	6.40 (33)	5.92 (42)	4.48 (49)	2.72 (49)	1.60 (91)
Mexico	10.41 (54)	9.48 (67)	4.62 (51)	2.89 (52)	1.73 (98)
Switzerland	7.00 (36)	5.60 (40)	4.20 (46)	3.28 (59)	1.40 (80)
Spain	2.54 (13)	1.78 (12)	1.27 (14)	0.89 (16)	0.63 (26)

Source: Arrillaga, 1955b, pp. 285–286.

Thanks to this progressive increase in foreign tourists in Spain, since 1950, tourism has been one of the most important factors to compensate for trade imbalances. According to an analyst of the time, these revenues were "like a golden rainfall which fertilises the income and gives strength to the national economy" (*Aragón*, 1955, nº 236, p. 4). Thus, by 1958, tourism revenues already accounted for a third of imports and more than half of the value of exports, thus contributing to alleviate the chronic trade balance deficit. Due to its

Tab. 6: Prices in hotels per person, 1936–1950 (current pesetas and by categories)

Year	Luxury	First A	First B	Second	Third
1936	45	25	20	13	6
1940	45	25	20	13	6
1941	60	35	25	15	10
1942	60	35	25	15	10
1943	60	35	25	15	10
1944	60	35	25	15	10
1945	60	35	25	15	10
1946	60	35	25	15	10
1947	60	35	25	15	10
1948	78	45	35	22.5	16
1949	100	60	40	25	20
1950	100	70	50	35	25

Source: Escorihuela, 1954, p. 50.

service exports, Spain not only managed to finance part of its trade deficit, but also to increase the amount of foreign exchange to import both raw materials and machinery and equipment goods for industry. Tourism had ceased to be a minor activity to become the sector that most drove the economy in those years (Arespacochaga, 1967, pp. 49–60; Larrinaga, 2016). (See Tab. 4.)

Having said that, we should, however, make a distinction between the different coastal destinations in Spain after the Civil War. Specifically, we are going to focus in the Balearic Islands, especially in Mallorca; and in a destinations of the Mediterranean coast, the Costa Brava in Catalonia.

3 The Balearic Islands. The case of Mallorca

In the years before the Civil War, the Balearic Islands (with Palma de Mallorca in top position), Barcelona, Guipúzcoa (with San Sebastian in front), Madrid and Seville led the tourist hierarchy of Spain from the point of view of receptive tourism. This means that they were the provinces that received the most foreigners. On the one hand, two large cities stand out, such as Madrid (the country's capital) and Barcelona (the venue, moreover, of the International Exhibition of 1929) and the most populated city in Andalusia, Seville, famous for its monuments and for having been the venue of the Ibero-American Exhibition of the same year. On the other, two important tourist destinations. One very consolidated, San Sebastian, a leading centre for French tourists, with France being the most important market of origin; and another, Mallorca, which emerged in the thirties to the international tourism market, by 1934 turned into an important destination for tourist flows to the Mediterranean (Vallejo, 2008, p. 142). In fact, taking into account the evolution of tourists staying in hotels in Mallorca, it went from 20,168 in 1930 to 40,045 in 1935, which is almost double the figure (Buades, 2004). Nor should we forget the significance of cruises, whereby Palma de Mallorca was a very interesting destination in the western Mediterranean (Cerchiello, 2017, pp. 95–98). Moreover, the turnover was around 30 million pesetas in 1934 which is an amazing figure at the time and makes us think that tourism played an anticyclical role in the economic crisis after 1929, characterized by a very considerable decrease in foreign trade. Moreover, it even served to promote a process of outsourcing that had begun in the previous decade (Barceló, 1966).

With the start of the Civil War, with the exception of some cases, most of the foreign holidaymakers in the Balearic Islands and on the Mediterranean or Atlantic coast left Spain by their own means or were evacuated by boat in their respective countries. This is what happened, for example, in the cases of Gerona and the Balearic Islands, where the British army played an important role (Vallejo, 2019a, p. 108). As expected, the war had a very strong impact on the Balearic Islands, due, fundamentally, to its island character (the isolation factor) and the high number of foreigners who already visited it in the thirties: 60 % of the total in 1933 and 52 % the following year (Cirer, 2009, p 268).

But little could be expected either of Spanish tourists, immersed in the conflict and with very restricted displacements. Consequently, the 1936 military uprising caused Mallorca's tourist activity to sink. Thus, a symptom of such a crisis was the closure of the emblematic Gran Hotel, which had opened its doors in 1903 (Vallejo, 2019a, pp. 110 & 119). On the other hand, the Promotion of Tourism in Mallorca (*Fomento del Turismo de Mallorca*), unlike other active tourism associations which also disappeared, managed to survive, despite a clear decrease in income (due to the fall in economic income from membership fees or institutional and hotel subsidies) and its partners, even some of them being shot. Therefore it is not surprising that, in reality, it led a languid life during the war years, waiting for better times (Vives, 2005, pp. 171–173). If truth be told, during these years, the island's main function was to serve as the centre of operations of the Italian air force in its efforts to bomb Barcelona and other republican towns (Buswell, 2011, p. 48).

During the Second World War there was a problem with communication. The only commercial companies which maintained regular air routes with Spain were the German Lufthansa and the Italian Ala Littoria. In 1939, the latter had a daily link with the Cádiz, Melilla, Palma and Rome route. In May of the following year, routes were introduced with Madrid and Barcelona and there was a daily flight between Rome and Palma. However, as the conflict progressed, the frequency decreased and all publicity disappeared for security reasons. By late 1944, all regular flights had been interrupted. As for internal flights, Franco's government made Seville, Salamanca and Zaragoza the basis of air communication in Spain during the first years of the war. Thus, the first operational line with Palma was from Zaragoza, practically irrelevant from the tourist point of view. Later the Madrid, Barcelona and Palma route would open, operated by Iberia. As for maritime communication, Franco's new regime gave strategic importance to improving connections with the archipelagos, both the Canary Islands and the Balearic Islands. In this way, Palma benefited from a link with Barcelona three times a week and with Mahon, Ibiza and Alicante once a week. Obviously, the scarce maritime links had a negative impact on the tourist flow. However, in 1942 four travel agencies had offices in Palma, namely: Wagons-Lits-Cook, Marsans, Viajes Iberia and Bakumar. Actually, the tourist engine of these years was the national clientele. The number of tourists staying in hotels in the Balearic Islands went from 40,045 in 1935 to 55,134 in 1945, the vast majority of whom were Spanish, since only 691 were foreign visitors in 1945 (Buades, 2004). Clearly, therefore, whilst waiting for economic conditions to improve in post-war Europe, national tourism reactivated the tourist flow of the islands. However, as early as 1949, the British Workers Travel Association, which had already included Mallorca in its itineraries in the thirties, re-incorporated it in 1949 (Buswell, 2011, p. 51). Without a doubt, it was a first step towards the recovery of international tourism on the island.

In this rather optimistic context, in January 1951, the First Tourist Assembly of Mallorca was held, promoted by the main tourist agents of the island, with the aim of promoting the sector and proposing actions such as the construction of a large hotel, a mountain hostel, a hospitality school and price control to curb inflation. The assembly served, then, to see the needs of the sector and to try to place Mallorca in the international tourism market. In fact, in June of that same year, the Council of Ministers approved the expansion of Palma airport, while authorizing its opening to international transit. This enabled companies such as Air France or British European Airways to be installed there, making Palma airport the third largest in Spain as regards volume of passengers by 1952 (Buades, 2004, pp. 128–129).

With such a different panorama, in January 1953, the Second Tourist Assembly of Mallorca was held. In it, a new financing system for the Promotion of Tourism in Mallorca was agreed based on the cooperation of the collaborating town councils so that the entity had 250,000 pesetas per year for broadcast and filmed tourist propaganda. A request was also made to remove visas for tourists from allies. Once communication had improved, especially air communication, it was essential to advertise the island in the markets of origin. Hence the need to intensify propaganda to attract foreign tourists. At this point, collaboration with travel agencies could be decisive. In this regard, it should be noted that in November 1953, the International Congress of Skål Clubs was held in Palma, a network of tourism-loving clubs that had been federated in 1934. Finally, in 1955 the XX Assembly also took place in Palma of the Spanish Federation of Tourism Initiative Centres (Buades, 2004, p. 133).

In short, all these meetings served to arouse the interest of the different tourist agents on the island of Mallorca. This was reflected in the arrival of tourists. Thus, as far as the number of foreign tourists was concerned, it rose from 31,600 in 1950 to 161,170 in 1956. Assuming 6.9 % of the total number of foreign tourists entered in Spain accounted for 10.3 % (Buades, 2004, p. 132). Such an increase in foreign tourists meant that in 1950 Spanish tourists staying in hotels accounted for 67.82 % of the total, six years later their share had dropped to 27.48 % (Fernández Fúster, 1991a, p. 443). The arrival of foreigners became the fundamental core of tourism in Mallorca. Tourism which, in the 1950s, became a true mass consumption industry. Luckily, if we look at the growth in the offer of tourist accommodation establishments, we see that it rose from 174 registered in 1950 to 235 in 1955. As for the number of hotel beds, the increase was even greater, practically 50 %, increasing from 4,054 to 6,828 (Buades, 2004, p. 130).

Despite the distance and clearly on that periphery of pleasure defined by Turner and Ash (1991), Mallorca became a preferred tourist destination in these years. Thanks to jet aeroplanes which meant you could arrive at your destination quickly, Mallorca became a fundamental enclave of international mass tourism (Buades, 2004, p. 137). In this regard, we must not forget that British businessmen were among the most active when it came to acquiring aeroplanes that had taken part in the war and converting them for civil use (Woodley, 2016). Among the most important companies we should highlight Horizon Holidays, founded by Vladimir Raitz, and Pontinental, founded by Fred Pontin, both very interested in the transfer of tourists to Mallorca (Buswell, 2011, pp. 51–52). In fact, Horizon Holidays began its trips to the island in 1952 (Bray & Raitz, 2001). (See Tab. 7.)

Tab. 7: Foreign tourists in Spain and the Balearic Islands, 1950–1956

Year	In Spain	In the Balearic Islands	% on all Spain
1950	456,968	31,600	6.90
1951	676,255	60,300	8.90
1952	776,820	72,100	9.20
1953	909,344	82,520	9.07
1954	993,100	88,740	8.90
1955	1,383,359	130,780	9.45
1956	1,560,856	161,170	10.30

Note: these are foreigners with passports.
Source: Buades, 2004, p. 132.

4 The conquest of the Mediterranean peninsular coast. The Costa Brava case

Both San Sebastian and the rest of the Atlantic coast of northern Spain and Galicia had to face the competition from the Mediterranean's new coastal destinations. In fact, there was a history, since it is true that some Mediterranean towns had become important tourist centres since the mid-19th century. They were, mainly, winter seasons, as the well-known case of the *Côte d'Azur* in France (Boyer, 2009). As for Spain, there are examples of sea water bathhouses in Barcelona from the end of the 18th century (Tatjer, 2012), although it was in the 19th century when there was an expansion of the installation of these bathhouses and the enhancement of warm beaches (Tatjer, 2018). However, these beaches did not get to enjoy the refined nature of the beaches of the northern peninsula during the first decades of the 20th century and even during the Franco regime. However, although with the precedents of the previous decades, it was after World War II that the great tourist flow to the Mediterranean coast occurred (Segreto, Manera & Pohl (Eds.), 2009).

For this first great tourist flow, a strong development of the automobile was initially necessary, which allowed numerous Europeans from the centre and the north to access the beaches of the south. For origin markets further away, the expansion of commercial aviation was of vital importance (Woodley, 2016). In fact, with the arrival of charter flights, the number of such tourists increased. Therefore, in this development of Mediterranean tourism, two moments can be observed, the first marked by cars, roads and highways and the second by charter flights. In this sense, charter aviation took place in airports in countries where car tourism had developed a sufficient hotel industry, namely where there was enough hotel infrastructure to accommodate a growing number of visitors. The case of southern France, Italy and Spain were paradigmatic (Fernández Fúster, 1991b, p. 595).

Having presented the general framework, it is worth emphasizing one of the favourite coastal destinations of these years: The Costa Brava. Initially, this coastal area extended from the north of Barcelona to the French border and was given this name due to the characteristics of its coastline. As in other parts of the Catalan coast, the first news about bathhouses in this area date back to the second half of the 19th century, although in very modest terms. In fact, it was only the arrival of the railway which began to enhance this area. We are talking about places of evasion of certain wealthy owners. So until the First World War, the Costa Brava was virtually unknown. This was not the case later on. We know that colonies of Belgians, Dutch, British and Germans (many of whom had escaped from Nazism) settled there in the early 1930s. Some even managed to open tourism businesses. This was the case of Nancy Johnstone and her husband Archie, who, in 1934, decided to leave London and set up a small hotel in Tossa de Mar. He was a journalist for the *News Chronicle*, a newspaper of the British liberals at the time, and all we know about her is that she was a great fan of the arts. Well, despite the outbreak of war in 1936, they remained in their establishment until 1939, when they ended up fleeing to France taking with them about thirty refugee children. His testimony, collected in two books (*Hotel in Spain*, 1937, and *Hotel in Flight*, 1939), can be an illustrative example of the diffuse foreign "touristification" characteristic not only of the Costa Brava, but also of other provinces of the Spanish Mediterranean in the 1930s (Vallejo, 2018, p. 150).

All in all, they had to wait for the Civil War and World War II to end for this coastal destination to gain more and more weight within the tourist destinations of Spain. In fact, in the forties, the problem of certain isolation continued since the main railway line went through inland, not along the coast, and the roads still left much to be desired. The economic troubles of the regime in its early years caused many of these road infrastructures to move into the background. However, there were many excursions and trips organized by coach from Barcelona, for example, sometimes extending on to Andorra, in these early years of the 1940s (*Hoja del lunes* (Barcelona), 12 July 1943, p. 6, announcement of Viajes Internacional Expreso). For foreign tourists, however, there was nothing odd about flying to Perpignan in France, to then travel by coach, cross the border and go to their summer destinations. So much so that, in the fifties, a large industry of tourist transport to the Costa Brava was set up in this area (Fernández Fúster, 1991a, p. 505).

Tab. 8: Amounts granted by Hotel Credit until 1957 in the main Spanish provinces (in thousands of pesetas)

Province	Total	The Capital of the Province	Rest of the Province
Gerona	66,710	-	66,710
Barcelona	55,800	46,900	8,900
Alicante	36,060	19,150	16,910
Malaga	35,350	1,250	34,100
Madrid	16,375	10,800	5,575
Valencia	15,750	11,750	4,000
Santander	15,500	9,650	5,850
Córdoba	15,000	15,000	-
Álava	14,000	14,000	-
Cáceres	13,588	7,588	6,000
Pontevedra	12,650	750	11,900
Valladolid	11,000	11,000	-
Cádiz	10,150	3,600	6,550
Oviedo	10,100	-	10,100
Guipúzcoa	9,500	-	9,500

Source: Aguirre, 1963, p. 113.

The increasing number of tourists arriving was also reflected in an increase in hotel establishments, in response to the shortage of hoteliers that some foreign travel agents had already noticed (*Hoja del lunes* (Barcelona), 16 March 1953, p. 3). Thus, according to an expert on the subject of the time, on the Costa Brava it rose from 119 lodging companies in 1955 to 437 in 1960, announcing in March 1962 that more than 130 hotels were being built in the area, some of them with foreign capital (Aguirre, 1963, pp. 110–111). Also, until 1957, the province of Gerona, to which a good part of the Costa Brava belonged, benefited the most from Hotel Credit. In collaboration with the General Directorate of Tourism, this was a specialized Service of the Industrial Credit Bank aimed precisely at promoting hotel development by supporting its financing (Brú, 1964, p. 6). The fact that 100 % appears invested in the rest of the province and not in the capital is a good indication that most of the investment was undoubtedly made precisely on the Costa Brava. In fact, it was not until 1964, once the Registry of Geotourism Denominations was created, that the exclusive identification of the Costa Brava with the coast of the province of Gerona occurred, thus avoiding any confusion (*Boletín Oficial del Estado* [*Official Spanish Gazette*], 26 December 1964, p. 17.311). In any case, the predominance of the investments of Hotel Credit can be seen in the provinces near the Mediterranean Sea (Gerona, Barcelona, Alicante and Malaga) (see Tab. 8), a faithful reflection of how much was happening in those years with the tourist flow.

5 Conclusions

The origins of coastal tourism in Spain date back to the second half of the 19th century, at least. Just as mineral waters were appreciated, so were marine waters. Different hygienist doctors also advised the intake of seawater as something healthy. Therefore, the first contact with the coast, from the tourist point of view, was related to hygiene and health. It was only at the end of that century, when medicine obtained more effective drugs than water intake, that the healthy character was replaced by an increasing sense of enjoyment. Hence, holidaying at the seaside was gaining importance among the most privileged groups of society. By then the cold waters of the European Atlantic façade had become fashionable, many times favoured by the presence of the different royal houses. Spain was no exception and its Atlantic coast became the favourite tourist destination of the wealthiest classes with San Sebastian and Santander at the helm. This predominance of this type of cold water remained until the late twenties and thirties, when certain water sports or the fashion of tanning began to gain more weight.

It was, however, after World War II, when the definitive transfer of tourist flows to the Mediterranean occurred, becoming the favourite destination of hundreds of thousands of holiday makers. Spain, which until then had occupied a lagging place in international tourism and had received a rather small number of foreign tourists, compared to France or especially Italy, began to see its beaches on the island of Mallorca or the Mediterranean side of its coast filled with tourists. The possibilities of cars and jet aircraft had a great impact on these movements. In addition, increasingly large groups of the European and even the American population joined the tourist practice, encouraged by economic well-being and the incentive of paid holidays. In this way, the choice of the Spanish Mediterranean coast guaranteed sunshine and warm weather at very competitive prices. Spain, within the span of a few decades, managed to rise to the peak of foreign tourists' favourite destinations with huge consequences, both from a social point of view, and, above all, tourism, due to its impact on the GDP. Tourist income also became one of the most significant compensatory items of its trade deficit. Hence some authors talk about a real Spanish tourist miracle during the fifties.

References

Aguirre, R. (1963). *El turismo en Guipúzcoa*. San Sebastián: Diputación de Guipúzcoa.

Arespacochaga, J. (1967). *Turismo y desarrollo*. Madrid: Servicio Informativo Español.

Arriba, E. (ed.) (1952). *Spain and Portugal in 1952*. The Hague: Travel Publications, Inc.

Arrillaga, J. I. (1955a). *El turismo en la economía nacional*. Madrid: Editora Nacional.

Arrillaga, J. I. (1955b). *Sistema de política turística*. Madrid: Aguilar.

Banco de Vizcaya (1960). "El turismo en España." *Revista financiera* (85), 11–19.

Barceló, B. (1966). "El turismo en Mallorca en la época de 1925–1936." *Boletín de la Cámara de Comercio, Industria y Navegación de Palma de Mallorca* (651-652), 47–61.

Barke, M. & Towner, J. (1996). "Exploring the History of Leisure and Tourism in Spain." In M. Barke, J. Towner & M. T. Newton (Eds.), *Tourism in Spain: critical issues* (pp. 3–35). Wallingford: CAB International.

Barke, M., Towner, J. & Newton, M.T. (Eds.) (1996). *Tourism in Spain: critical issues*. Wallingford: CAB International.

Boyer, M. (2002). "El turismo en Europa, de la Edad Moderna al siglo XX." *Historia Contemporánea* (25), 13–31.

Boyer, M. (2009). *L'hiver dans le Midi. L'invention de la Côte d'Azur, XVIIIe-XXIe siècle*. Paris: L'Harmattan.

Bray, R. & Raitz, V. (2001). *Flight to the Sun*. London: Continuum.

Brú, J. (1964). *El Crédito Hotelero en España*. Madrid: Instituto de Estudios Turísticos.

Buades, J. (2004). *On brilla el sol*. Eivissa: Rex Publica.

Cerchiello, G. (2017). *La evolución de los cruceros marítimos en España*. Valencia: Universitat de València.

Cirer, J. C. (2009). *La invenció del turismo a Mallorca*. Palma: Documenta Balear.

Correyero, B. & Cal, R. (2008). *Turismo: la mayor propaganda de Estado*. Madrid: Vision Net.

Escorihuela, E. (1954). "Los precios de la industria hotelera." *Revista Sindical de Estadística* (36), pp. 49–54.

Fernández Fúster, L. (1991a). *Geografía general del turismo de masas*. Madrid: Alianza.

Fernández Fúster, L. (1991b). *Historia general del turismo de masas*. Madrid: Alianza.

Fernández Navarrete, D. (2005). "La política económica exterior del franquismo: del aislamiento a la apertura." *Historia Contemporánea* (30), 49-78.

Fodor, E. (1952). *Spain and Portugal in 1952*. The Hague: Travel Publications, Inc.

Gil de Arriba, C. (2018). "Los espacios litorales españoles en la estructuración de las geografías turísticas del primer tercio del siglo XX." In R. Vallejo & C. Larrinaga (Dirs.), *Los orígenes del turismo moderno en España* (pp. 171-211). Madrid: Sílex.

Gil de Arriba, C. & Larrinaga, C. (2021). "La cornisa cantábrica como región turística en las primeras décadas del siglo XX (1902-1931)." *Investigaciones de Historia Económica* (17-1), 26-36.

Larrinaga, C. (2015). "De las playas frías a las playas templadas: la popularización del turismo de ola en España en el siglo XX." *Cuadernos de Historia Contemporánea* (37), 67-87.

Larrinaga, C. (2016). "El impacto económico del turismo receptivo en España en el siglo XX (1900-1975)." *Revista de la historia de la economía y de la empresa* (X), 23-50.

Lickorish, L. J. & Kershaw, A. G. (1958). *The Travel Trade*. London: Practical Press Ltd.

Norval, A. J. (1936). *The Tourist Industry*. London: Sir Isaac Pitman & Sons.

Pack, S. D. (2006). *Tourism and Dictatorship. Europe's Peaceful Invasion of Franco's Spain*. London: Palgrave Macmillan.

Pellejero, C. (2005). "Turismo y Economía en la Málaga del siglo XX." *Revista de Historia Industrial* (29), 87-114.

Pemble, J. (1987). *The Mediterranean Passion*. Oxford: Clarendon Press.

Segreto, L., Manera, C. & Pohl, M. (Eds) (2009). *Europe at the seaside*. New York & Oxford: Berghahn Books.

Soane, J.V.N. (1993). *Fashionable resort regions*. Wallingford: CAB International.

Tatjer, M. (2012). *Els banys de mar a Catalunya*. Barcelona: Albertí.

Tatjer, M. (2018). "Los balnearios catalanes y el contexto peninsular. Historiografía e historia." In J. M. Puigvert & N. Figueras (Coords.), *Balnearios, veraneo, literatura* (pp. 25-71). Madrid: Marcial Pons.

Turner, L. & Ash, J. (1991). *The Golden Hordes*. London: Constable.

Vallejo, R. (2013). "Turismo y desarrollo en España durante el franquismo, 1939-1975," *Revista de la historia de la economía y de la empresa* (7), pp. 423-452.

Vallejo, R. (2018). "La formación de un sistema turístico nacional con variaciones regionales." In R. Vallejo & C. Larrinaga (Dirs.), *Los orígenes del turismo moderno en España* (pp. 67–170). Madrid: Sílex.

Vallejo, R. (2019a). "Turismo durante la Guerra Civil, 1936-1939." *Revista de Historia Industrial* (75), pp. 97–132.

Vallejo, R. (2019b). "Turismo en España durante el primer tercio del siglo XX: la conformación de un sistema turístico." *Ayer* (114), pp. 175–211.

Woodley, C. (2016). *Flying to the Sun*. Stroud: The History Press.

Marta Luque and Víctor M. Heredia[2]

Pioneering projects in the tourism development of the Costa del Sol (Spain)[1]

Abstract The Costa del Sol is one of the most important areas of Spanish tourism development in the last century. Some of its municipalities have been a major focus of European mass tourism since the 1960s. In this work several official and private projects are proposed that committed then to the creation of a nucleus of health and leisure tourism in Torremolinos during the first half of the 20th century.

1 Introduction

The origin of the Costa del Sol as a tourist destination must be placed in the early years of the 20th century. It is then when private and public initiatives begin to be launched, turning a very specific area of Malaga coastline into a world tourism benchmark. The capital of the province, and above all, some small fishing municipalities on its western coast, underwent a transformation in the first half of the 20th century based on the construction of a network of tourist services, which led them, from the decade of the fifties, to become international centres of tourist attraction. Some of these initiatives are the object of our study.

The birth of tourism on the Costa del Sol is closely linked to the foreign colony established in Malaga and its western coast. And although the activity carried out by these foreigners was very varied, as recorded in the lists of the different consulates, many of them chose Malaga for health reasons or just as a resting resort (Burgos Madroñero, 1974). At the beginning of the 20th century, Malaga was already receiving visitors due to its pleasant climate, particularly in winter, and those pioneers who foresaw the possibility of tourist development in Malaga and its western coast based on that very fact. The projects that we analyse in this chapter contributed, no doubt, to the tourist development of Costa del Sol: the opening of an aerodrome in 1919, served as a basis for the future massive arrival of tourists by air; the birth of the offer of tourist accommodation;

---

1 This study forms part of the research Project HAR2017-82679-C2-1-P, financed by the Minister of Science, Innovation and Universities of the Government of Spain and the ERDF.
2 The authors want to express their gratitude to Mr. Enrique Girón for his help.

the construction of a golf course; and, an ambitious business initiative, which, although frustrated its initial plans, intended, already in 1935, to turn a Costa del Sol town, Torremolinos, into an important avant-garde tourist resort.

2 *El Rompedizo* Aerodrome, an infrastructure for the future

In the early years of the 20th century, a transport infrastructure was born, though with a very limited use, but it became afterwards essential for the tourist development of the Costa del Sol. The birth of the Malaga airport was largely accidental, a response to the need to have spaces so that the incipient aircrafts could make refueling and provisioning stops on the first commercial routes.

When in 1918 a Frenchman named Pierre Latécoère founded one of the first postal airlines with the intention of establishing a line between Paris and Buenos Aires, he chose Malaga as a support station. The first flight arrived in Malaga on 9 May 1919, and as the previously chosen landing field got flooded due to the heavy rain, the pilot had to find elsewhere to land. He found it at the *El Rompedizo* farm, located next to the road that connects Malaga and Torremolinos. The airline immediately rented the farm, thus becoming the first stable aerodrome in Malaga. Once the government permits were obtained to fly over Spanish territory, the first regular flight was made on 1 September 1919. The flights had a periodicity of twice a week and although the main object of the transport was mail and certain merchandise, at the end of 1920 the airline had already transported 203 passengers (Utrilla Navarro, 1994, pp. 53–62).

In the following years, the airline carried out land levelling, pulling up weeds, construction of a hangar and elements to facilitate landing and takeoff operations. In 1924 a radio station was installed and the following year a beacon system made up of electric bulbs, which marked the area of the airfield. Although the number of travellers was very small, the company had a vehicle to transfer them to their accommodation in Malaga or to other nearby towns. One of the first famous passengers to pass through the Malaga airfield was King Albert of Belgium, on his trip to Morocco in October 1921. Among the company's personnel, who were all French nationals, was the pilot Antoine de Saint-Exupéry, famous writer (Utrilla Navarro, 1994).

In 1926 the magazine published by the British colony in Malaga, *British Colony Gazette*, exposed the different ways to get to the city at that time. He considered that the easiest and most used way was by sea to Gibraltar, on a three or four-day voyage. From Gibraltar, travellers could rent a car to travel the 70 miles that separated the British colony from Malaga, although there was also the possibility of going by bus. The rail journey from England was shorter if it was done directly, but they advised a stay of several days in Madrid. Finally, the magazine pointed out the alternative of the future: "But the shortest route is by air, to Paris, an express train at night to Toulouse, and then by Lignes Latécoère (airplane) to Malaga, whose journey lasts 27 hours from London to Malaga." The use of the plane gave the opportunity to carry ten kilos of luggage and check the rest in a steamer (*British Colony Gazette*, 21 October 1926, p.5).

Primo de Rivera's government created in 1927 the Higher Council of Aeronautics and on July 19 of that year a decree-law was published that laid the foundations for the construction and operation of a network of national airports. The execution of works and the management of the airports was in the hands of some organizations that would be created with a specific nature, inspired by the Port Works Board. In January 1928 a first air communication plan was published declaring the Barcelona-Valencia-Malaga-Seville and Melilla-Malaga-Ceuta-Cadiz lines of public interest. The Malaga National Airport Board was set up, chaired by Luis Fernández de Villavicencio and made up of authorities and representatives of local and provincial institutions. The Technical Commission in charge of informing about the appropriate spaces for the location of the airports came to the conclusion that in Malaga the most suitable site was the *El Rompedizo* farm itself. Thus, it was decided in 1928 to carry out the acquisition of the land and extend it with parts of neighbouring farms in order to have a sufficient surface to convert it into a civil and military airport. The purchase by the State was not completed until the publication of a decree to that effect signed on 9 March 1932 (Utrilla Navarro, 1994, pp. 103–111).

The growing distrust of the Spanish government towards the flights of the French company, which had been absorbed in 1927 by the French State to create the Compagnie Générale Aéropostale, in a sector that was beginning to be recognized as strategic and in which it was committed to the creation of state airlines, forced it to stop operating in Malaga in the 1930s. Meanwhile, the *El Rompedizo* aerodrome was also used for sporting events such as the Air Tour of Andalusia in 1929 or the Tour of Spain in 1931. In this last year Málaga aeroclub was created, which would be in charge of organizing various aeronautical festivals before the Civil War.

The works to provide the airport with basic infrastructures that would allow it to provide services for regular lines continued until after the Civil War. During the conflict it was enabled as an air base and it was shelled and its facilities were damaged. When Franco's troops occupied Malaga and took over the airfield, repair and expansion works were undertaken, both on the runway and on the facilities. Several buildings were built, including the first control tower, inaugurated in 1938, and a passenger station for civil aviation, both designed by the architect Luis Gutiérrez de Soto (Aguado and Utrilla, 2010, pp. 6-8). Commercial flights were resumed by the Italian company Ala Littoria, which established routes with Tetouan, Melilla and Seville, being able to link from the latter city with Rome (Utrilla Navarro, 1998, pp. 52-53).

Malaga City Council assumed the cost of the new passenger station to ensure the use of the airport for civil flights. In March 1939, the awareness that it was an investment for the future was included in the municipal act: "A civil airport not only to strengthen current air communications, but to increase them, for the future and consequently the intensification tourism, which will bring immeasurable benefits for Malaga." That same year, the City Council acquired several neighbouring farms in order to expand the surface of the aerodrome and agreed to put the name of the aviator García Morato to the civil airport. The works of the passenger station, projected as a large building open to the outside through arches and a terrace, lasted until 1948. By then the airfield had been conditioned and *El Rompedizo* had been recognized in 1946 as a customs airfield open to international traffic. Although this measure barely had immediate effects due to the isolation suffered by the country and the low activity of the airlines (Utrilla Navarro, 1994, pp. 131-138).

The only company that used the terminal after Ala Littoria disappeared was Iberia (first as Spanish Air Traffic), which maintained air services with Madrid, Melilla and Tetouan from May 1940, later extended with the Seville-Malaga-Melilla line. For non-scheduled flights the activity of the Compañía Auxiliar de Navegación Aérea (CANA) was authorized, which had two nine-seater aircraft to make the Tetouan-Melilla-Malaga-Granada-Malaga-Tetouan route, which enjoyed high demand despite its limited capacity. CANA operated between 1947 and 1949. For its part, Aviaco made flights since 1948 with Seville and Barcelona, and Gibraltar Airways established lines between the colony and Tetouan, Seville and Malaga, operating this last line for just three months in 1950 (Utrilla Navarro, 1994, pp. 121-130). Gibraltar airport was built during the Second World War and after that it became a civilian airport, initially becoming the British gateway by air to the Andalusian coast.

It was at the end of the fifties when charter flights and foreign airlines began to operate linking the Costa del Sol with northern Europe. By then the foundations were already set up in a space that would prove fundamental for the future takeoff of mass tourism from the following decade. One of those bases was its location, halfway between Malaga and Torremolinos and next to the main road and the suburban railway line. A location that provided numerous advantages for the future mobility that would be generated around the airport.

3 The first accommodations for tourists

When the 20th century began, the only hotel establishments aimed at hosting tourists were in the capital of the province. In Malaga there were already some more than acceptable, with a pleasant atmosphere and in which mainly foreigners who came to the city to enjoy its pleasant weather, especially during the winter season, stayed. However, the towns on the western coast, so emblematic in today's world tourism, such as Torremolinos or Marbella, lacked a hotel offer capable of attracting tourists, they only had traditional inns and small urban hotels designed to house business travellers or occasional travellers, unpretentious and reluctant to spend too much. It was not until the first decades of the 20th century, when hotel establishments capable of meeting the needs of these new tourists, mainly European clients with high purchasing power, would appear along the coast. From the middle of the decade of 1920 those tourists seemed to have overcome the traumas and the effects of the Great War (Arenas and Majada, 2003). All these establishments were characterized by their small size, their almost exclusively foreign clientele (and for the most part of British origin) and a family service that offered certain comforts in addition to the benefits of a paradisaical coastline with an extremely excellent climate. Although there were other ones, both in the city of Malaga itself and elsewhere on the coast, we are going to focus on two pioneering hotels in Torremolinos and Marbella: Santa Clara and Miramar.

3.1 Santa Clara (Torremolinos)[3]

As we pointed out before, if the development of tourism on the Costa del Sol is highly linked to the colony of foreigners settled in Malaga, it is not strange that the one who holds the not inconsiderable title of being the first hotel on the Costa de Sol, is a British man, George Lanworthy. On the other hand, it is also reasonable that the first hotel was established on a recreational estate, since Torremolinos was a place chosen by well-to-do families to spend the summer period, which led to the appearance in this town of important villas and recreational estates.

---

3 The documentary sources used in this investigation are primary sources owned by the heirs of the workers of Santa Clara, Campoy and Salas, as well as those deposited in the Archivo Histórico Municipal de Antequera, Fondos Familiares (family archives) Archivo Ramos.

Santa Clara, later also known as El Castillo del Inglés, was one of the most important recreational estates in Torremolinos. Formed from the successive union of several pieces of lands, it included among them a fort located on the rocky promontory called Punta de Torremolinos, whose construction in 1763 was intended to protect the coast and serve as a refuge for vessels exposed to the action of corsair vessels who frequented this coastline. The small fortress passed into private hands in 1873, and after being the subject of successive purchases and sales, it was acquired, together with another adjacent property, by the English military man George Langworthy in 1905. At first the intention of the buyer was to lease the property, to which, two days after signing the purchase deed, put an ad in the Malaga newspaper *La Unión Mercantil*: "The magnificent Santa Clara hotel is rented with a good garden and several small detached houses […], with great comfort to take sea baths, as it is by the beach." However, and without knowing the reasons, it was Langworthy and his wife who settled in Santa Clara, appearing as residents in February 1906.

After the purchase, Langworthy conditioned the estate, kept the main house as a simple farmhouse and, as Luis Bello reported in the Madrid newspaper *El Sol*, he knew how to take advantage of the decorative elements of native nature, with viewpoints with rustic seats, paths between the rocks and with a boat stopover to access the beach (*El Sol*, 14 August 1926). Santa Clara became a tourist attraction where famous figures flocked to, such as in 1926 King Alfonso XIII (*El Cronista*, 16 February 1926, p. 4), or in 1927 Queen Victoria Eugenia and her son, the Prince of Asturias, who they came to have tea on its splendid balcony overlooking the sea (*El Telegrama del Rif*, 2 March 1927, p. 2). Besides the Sociedad Excursionista de Málaga included among its activities the visit to the Englishman's farm, because "from its beautiful viewpoints one can reach the marvellous view of the Malaga coast" (Sociedad Excursionista de Málaga, 1929, p. 92). In the *Guía del Turista. Andalucía*, 1929, in the excursion to Torremolinos it is recommended to visit the Castle of Santa Clara, for which the permission of the gardener was required (Koethke, 1929, p. 33). Gardeners in Santa Clara played an important role, since Langworthy, who had retired to a small house on his farm, gave them the exploitation of it.

The first reference to hotel activity in Santa Clara appeared in 1930 in an advertisement published in the weekly English news page of the newspaper *La Unión Mercantil*, "Santa Clara, Torremolinos. – English lady receives paying guests. Afternoon teas served. Tennis" (*La Unión Mercantil*, 11 June 1930). It was an English lady, Margaret Horn (Mrs. Beautell) who rented the farm with a partner for hotel use. Some of the buildings on the estate were dedicated to this new activity: the old police barracks, which underwent constant improvements, was used to house the rooms, while in the main house there were, in addition to the kitchen, the common areas, reception, dining room and lounges. Some other dwellings on the estate, owned by Langworthy employees, were also given hotel use. Most of the clientele was foreign, mainly British. In the following years the business was run by Nancy Beutell, Margaret's daughter, and her husband, the American Mark Hawker. On 29 February 1940, Antonio Campoy López, as a representative of the Santa Clara workers, sublet the hotel to Frederick Saunders, who ran it until 1957 when he was evicted for non-payment, at a time when the hotel had twenty-three employees.

3.2 Hotel Miramar (Marbella)[4]

In October 1933 the Miramar Hotel was inaugurated located in the Huerta de San Ramón, next to the urban area of Marbella, and run by the couple formed by José de Laguno Cañas and Agustina Zuzuarregui Sotto-Clonard. Before arriving in Marbella, José and Agustina had lived in Madrid, Cuba, Gijón and San Sebastián. For medical advice, due to the illness of one of their daughters, they changed the Cantabrian Sea for the Mediterranean, moving their residence initially to Valencia, and finally, in 1930, to Marbella. Upon reaching the city, they acquired the Huerta de San Ramón, a farm located between the road (which had just been paved) and the beach, with a vineyard, grove and orchard. There was a farmhouse next to which they built a large villa to accommodate his large family. Although at first, they thought to live off the farm's agricultural production, the poor economic results led them to undertake a different business adventure: the opening of a hotel. Surely this decision was made influenced by José Zuzuarregui, Agustina's brother, lawyer, councillor and promoter of the benefits of the Marbella coastline, who a year before the hotel was opened published an article in the Malaga newspaper, *El Cronista*, which under the name of *Marbella*, praised the benefits of this town, and pointed out that it was destined to be one of the first centres of world tourism, for which it had to develop a hotel industry (*El Cronista*, 1 December 1932, p. 1).

After making the necessary reforms, the family settled in the original house in the garden and the villa was transformed into a small accommodation with about ten or twelve rooms that was inaugurated on 15 October 1933. The press that published the news of the opening highlighted the attendance of the local

---

[4] The information in this section has been provided, for the most part, by Mrs. Josefina de Laguno, whom we thank for her kindness.

authorities and the excellent location of the new establishment, close to the town centre and bordering the best beach for bathing in Marbella. The equipment of the hotel was to be completed with a lush park, a tennis court and a parking lot. The correspondent stated with clear judgment: "Marbella is still the unknown pearl. It has been a great success to promote the hotel industry and thanks to it tourism in the small city, opening channels of progress for its magnificent tourism future." On the terrace of the hotel, which enjoyed magnificent views over the coastline, it appeared in luminous letters: El Miramar (*El Cronista*, 20 October 1933, p. 1).

The new establishment was registered as a luxury hotel and carried out an extensive advertising campaign in the Anglo-Saxon and Moroccan media (*Le Petit Marocain*, 20 May 1936). The list of services offered and their prices were collected in a small brochure published in English and French, full board amounted to 15 pesetas a day. Its clientele came mainly from Gibraltar and the most important hotels in Malaga, such as the Miramar or the Caleta Palace. They housed groups of students from Oxford and Cambridge in the winter season and also received Spanish guests such as Bernabé Fernández Sánchez, creator of Ceregumil (a well-known dietary supplement), Norberto Goizueta, who in 1934 acquired the Hacienda Guadalmina, or Alberto Insúa, Hispanic-Cuban writer and politician, and a good connoisseur of the excellencies of the Marbella coast, which he already called Costabella in the late 1930s. One of the attractions of the place was the restaurant, with an international menu.

The promising future of the hotel, which was already using the sunny coast concept in its advertising, perhaps the first use of the Costa del Sol idea, was cut short by the tense social environment that the country was experiencing. During the initial months of the Civil War the family lived in a permanent state of siege, surviving thanks to the help they received from the fishermen in the area, grateful for the social work they had developed by opening an evening school. Mattresses and bedding were seized for the hospital and could not be recovered until January 1937. When the hotel reopened, the demand for accommodation was already very low. Only some families from the interior came to Marbella and the foreigners had disappeared with the war. The Laguno Zuzuarregui family finally made the decision to abandon this business venture, move to Malaga and sell the farm. In this way, the Civil War put an end to the brief hotel history of the Miramar.

4 The Golf Course[5]

Between Malaga and Torremolinos another basic infrastructure was created for the development of tourism on the Costa del Sol. And as with the airport, its genesis period was long and its effects were not evident until decades later, but the reality is that it was a project conceived and developed in the twenties and thirties of the 20th century. We mean the golf course.

Golf penetrated into Spain through the British colonies, which spread the love among the aristocracy and the bourgeoisie. In the case of Malaga, in the mid-1920s a group of local personalities linked to the Sindicato de Iniciativas de Málaga (Malaga Initiatives Union) proposed the construction of a golf course as "one of the most urgent needs to attract outsiders" (Luque Aranda, 2015, p. 63).

On 9 March 1926, twelve personalities from Malaga society met to establish a golf company, considering it necessary for the promotion of tourism and for Malaga in general, thus creating the Malaga Golf Club, a sports association, initially chaired then by the mayor of the city, José Gálvez Ginachero, from whom the idea of establishing it had arisen. To obtain the capital necessary for the installation of the golf course, a bond issue was carried out. The registered office was located in a premise owned by the partner Luis Fernández de Villavicencio, who in 1927 succeeded Gálvez Ginachero as president. On 24 April 1926, the Civil Government approved the regulations of the company, although it was not legally constituted until December 17 of that same year. Taking advantage of the stays of the members of the royal family at the Hotel Príncipe de Asturias, the partners contacted Queen Victoria Eugenia and her mother, Princess Beatrice (daughter of Queen Victoria of the United Kingdom), who were interested in the project, considering that the possibility of practicing this sport would be a powerful attraction to attract winter tourism (Zarca, 1998, p. 84). In fact, the regulations indicate that the company was created in order to fill a need that had been vividly felt for years but that then was already urgent and unavoidable, and that if it were not realized, the hopes that Malaga would convert into an attractive winter season would be largely frustrated. The creation of a golf course would allow winter visitors to practice this sport, essential even then in tourist cities.

In January 1927, the members began to visit farms where they could build the golf course, a task that was joined in March by a professional from the Puerta de Hierro Club in Madrid, Mr. Gallarza, who was hired for 500 pesetas for this advice. At the beginning, it was also agreed that he would be the one

---

5 For the study of this project, we have used the primary sources deposited in the Archivo Histórico Municipal de Antequera, Fondos Familiares, Archivo Ramos.

who would design the layout of the field, for which he would charge 1.200 pesetas, although in the end this task was not carried out. Of all the lands visited, they agree to buy a plot of land from the Cortijo de Velarde, part of the Hacienda Nueva Colonia located between Malaga and Torremolinos, very close to the mouth of the Guadalhorce River and from the sea, because it has a very suitable configuration for golf use, with an area of 44 hectares. In March 1928, Queen Victoria Eugenia gave the president of Malaga Golf Club a check for 20.000 pesetas that the City Council had consigned to subsidize the golf course (Zarca, 1998, p. 84). At that time, the land purchase contract was signed before a notary, advancing the owners 28.000 pesetas of the agreed price, which amounted to 55.000 pesetas. On April 28, Mr. Colt, a British expert in the field, proposed to design the course of the field, who came to Spain to do the same in a field in Madrid and another in Santander, so the travel expenses would be borne by the three. His fees amounted to 250 guineas and in November of that same year the project carried out was already presented at the Board of the company. Seeing that the resources that the Malaga entities could contribute were not enough during that summer, negotiations were made with the governing body of tourism at the national level, Patronato Nacional de Turismo (PNT) (the National Tourism Board), to obtain material support that would allow the construction of the field. The PNT granted a loan of 425.000 pesetas for the construction of a golf course and a clubhouse.

The works began in 1929 (*La Unión Mercantil*, 20 January 1929, p.2). The PNT was advancing amounts of the loan granted, which allowed the company to deed the purchase of the land on 25 November 1929.

With full control of the land, on 28 February 1930, the mortgage loan granted by the PNT to the Málaga Golf Club company was formalized before a notary public for an amount of 425.000 pesetas, at an interest of 5 %, an amount of which had been anticipated 125.000 pesetas. The repayment of the principal and payment of interest would be made by delivering at the end of each year all the income generated by the company, discounting general expenses. It was agreed that while the mortgage was maintained, the PNT would intervene in the regime and administration of the club with two delegates. In addition, the company was obliged to prepare the field and build the clubhouse within a period not exceeding 20 months from the date of the signing of the mortgage, keep what was built, insure it against fires and pay taxes. In the following days of the signing of the mortgage, the representatives of the company address the delegate of the PNT in Andalusia, Luis Bolín, to request him to hand over them the sum of 300.000 pesetas that remained unpaid.

On 16 November 1930, and after an inspection visit by members of the PNT to the works of the field, Bolín met with representatives of the society, and it was agreed that, in view of the excellent conditions of the field, to give greater amplitude to meet the most demanding demands, and make it the best golf course in Europe. The company estimated the necessary budget for this and requested a subsidy of 275.000 pesetas from the PNT, which was granted on the 24th of the same month. It was Bolín who communicated the decision of the public body explaining that it was based on the extraordinary tourist interest of the Golf Club of Malaga, as well as how convenient it would be for the city to have such a valuable tourist attraction.

The political change (the arrival of the 2nd Republic) paralyzed construction because, although in October 1931 the field was practically finished, save the building of the club house, some complementary works and the completion of the access road from the Cádiz highway, of the 275.000 pesetas promised only had been received 25.000. The PNT was not willing to complete the rest of the promised funds so it was impossible to finish the field. But given that in the opinion of the partners, what interested Malaga was that it be finished so that they could play as soon as possible and thus safeguard the interests of the city, they offered all their facilities, equipment, supplies and everything that belonged to the PNT. The offer of the transfer of the property was without limitations or reserves in exchange for the company being released from mortgage liability. The PNT took possession of the golf course in November 1932, that same year the works were resumed, being the golf course finally inaugurated in July 1934 with the assistance of the civil governor, Alberto Insúa (*Mundo Gráfico*, 11 July 1934, p. 29). It came into operation with nine holes to play in the summer and another nine as a winter round, when it was fully used. As a provisional clubhouse, the Refugio de Chochales was set up. In Los Pinos the technical service led by Professor Julio Casaña was installed, in addition to the sale and rental of bags, clubs and balls. The main building, the tennis and badminton courts and the spa pavilion were pending completion (*Blanco y Negro*, 2 September 1934, pp. 40–41).

Training and competitions were immediately scheduled and an agenda was drawn up for the incipient local fans and for the British tourists who opened the course, which was open all year and whose competitions were free for amateurs of both sexes. Even then it was known as the Torremolinos golf course (*ABC*, 19 December 1934, p. 56). During the Civil War the countryside was seriously damaged. Then came the reconstruction, expansion, construction of the inn and the parador and the founding of the Country Club.

5 Torremolinos S.A.[6]

The weather conditions of the Malaga coast met conditions similar, if not better, to those of the French Midi, which, from October to April, attracted the nobility and upper bourgeoisie from the last third of the 18th century. In the French Riviera, winter stations such as Cannes, Nice or Saint-Tropez multiplied, but they also emerged further east, on the Italian coast, and to the west, in some cities of the Spanish Mediterranean, such as Alicante, Palma (Majorca) or Malaga. With the arrival of the 20th century, the taste for the sun and swimming was born, and the vacationers who traditionally came to take the waters in the cold seas, began to move towards warm seas, so these winter seasons were transformed into summer seasons (Boyer, 2002). These transformations were mainly due to those with a visionary entrepreneurial attitude who joined forces to consolidate some of these towns as important tourist centres, and among these towns are some of those that make up the current Costa del Sol.

There was a very interesting initiative that intended to provide Torremolinos with a strong tourist structure, thus turning this town into an important tourism receiving centre. This project, designed by the architect Gonzalo Iglesias, was born in 1926. It consisted of a large, complete and ambitious project. In its launch report the exceptional conditions of Torremolinos and the need for their improvement were exposed. He pointed out the excellent conditions of this town so as to turn it into a winter resort and into a place to visit, as it is located only 12 kilometres from Malaga, a distance that could be covered in a pleasant car ride. Its assets include its exceptional climate, the exceptional beauty of the landscape, with a beautiful and extensive beach, its clean and healthy water, and cheerful and good-natured inhabitants. Iglesias defended that the permanent, what derives from the location and its natural conditions, was unbeatable; whereas the deficient was susceptible to easy and rapid improvement because it was accidental.

According to the author of the project, 1926 was an ideal moment to undertake the improvements of the deficiencies that had prevented Torremolinos from being the satellite city of Malaga and the ideal season for both winter and summer. To do this, he proposed to carry out an intense and organized action that would simultaneously develop improvements in various orders (urban planning, hygiene, supplies, spa, hotels, cleaning, sports, tourism, building, comfort, communications, attractions, order, etc.), to make Torremolinos the picturesque, clean, pleasant and beautiful town that it could and should be. The

---

6   For the study of this project, we have used the primary sources deposited in the Archivo Histórico Municipal de Antequera, Fondos Familiares, Archivo Ramos.

project required a capital investment that would not have compensation in the work itself -improvements in the cleaning and hygiene of the population- so that sufficient interest to stimulate capital would have to be sought in the increase in the value of the property. The way forward would be to establish a public limited company that would carry out a complete and organized action for the benefit of the population and obtain compensation in the increase in the value of the property, especially the land that would be incorporated into it. All this under the name of *Torremolinos S.A.*

The main support of Iglesias in this project was the Malaga lawyer Enrique Ramos Puente, who would contribute to it with a rustic farm of his property *Huerta de Cantó*, located in the La Carihuela neighbourhood. Although Ramos Puente and Iglesias agreed that the constitution of the company should be finalized before February 1928, failed to materialize, perhaps due to Iglesias´s overwork. Gonzalo Iglesias had presented a project for 1.500 homes in Seville that was approved in 1927. The initial destination of these homes was the accommodation of travellers during the Ibero-American Exhibition that was to be held in Seville in 1929. Once said contest was over, they would be used for the purposes prescribed in the Royal Decree of Cheap Houses of 10 October 1924. We gather that this work definitively removed Iglesias from his Torremolinos project, since at his death, on 23 June 1934, the Sevillian project had not finished (*ABC Seville*, 27 June 1934, p. 36). But the fact that Iglesias abandoned, or at least postponed, the project for the urban development of Torremolinos did not mean that his partner did it too. In the summer of 1931, Ramos Puente planned the construction of a beach resort located under the Santa Clara farm. To carry it out, it would constitute a cooperative of which the workers who registered to work and those who contributed the necessary cash would be a part. The works began with the construction of the descent from the Castillo del Inglés, but the project did not prosper (*Nuevo Mundo*, 11 September 1931, p. 35). However, the idea of the spa was maintained and evolved until it reached a more complex dimension.

In 1934 a report was presented on the Project for the Construction of a Spa in Torremolinos. In it, the exceptional conditions of this town were once again highlighted: location, ease of communication with the capital, orientation, beauty, climate, etc., which, even without stimulation or propagation, had been sufficient to attract a growing current of people who visited it and especially of foreigners who settled there. The report also affirmed that the only thing done successfully to take advantage of that current had been the work of foreigners themselves, except for the "magnificent" road that crossed Torremolinos and the construction in the vicinity of the golf course. All of this generated an excellent opportunity that should not be missed, given the certainty of success: there was scope for all kinds of initiatives, both the most audacious that would require large capital to tend to force development and progress, as well as the most modest ones, which would limit themselves to attend to already manifest needs, preventing the ongoing development from stagnating or paralyzing.

The business he was proposing would begin with the construction of a low-cost beach resort, susceptible to successive extensions. The site chosen was the magnificent and wide 70-metre-wide beach in front of the estates called *Cortijo del Tajo* and *La Gotera*. To this end, a purchase option contract had been obtained for the land required for the first and had the willingness of the owner of the second, Antonio Navajas Ruiz, representative of the Casa Larios, to participate in the business. *Villa Pepita* was located in *La Gotera*, the now emblematic house of the Navajas. The concession of the beach was planned to apply as soon as the entity to be constituted had legal personality. This beach had unbeatable conditions for the purpose it was pursuing: clean, with fine sand of a pleasant colour and with a very soft entry into the sea.

To carry out the project, a public limited company is set up under the trademark proposed by Gonzalo Iglesias, who died a month before the presentation of the aforementioned report. On 16 January 1935, Antonio Navajas Ruiz, Juan Fazio Cárdenas, Enrique Ramos Puente and Manuel Luque Lavado, the latter owner of the *Cortijo del Tajo*, formalized the deed of incorporation of the company, *Torremolinos S.A.* With a share capital of 500.000 pesetas and with the main purpose of building and operating a spa on the beaches of Torremolinos, the acquisition, sale, urbanization, parceling of land and construction of the same was added as a corporate purpose, either to dispose of them or to exploit them as a tourist rental, such as pensions, hotels, restaurants, entertainment venues or for sports practices. In short, any business that was related to the development of tourism, including the establishment of lines for transportation service.

That same year, 1935, the company was launched. The first works carried out were the enclosure (placement of posts and wire), and later the beautification of the land by planting trees and plants from a nursery in Valencia. In parallel and as a prerequisite to the installation of the spa, it was agreed with the City Council the construction of the access roads. Liquidity needs were increasing, so in 1936 a public subscription of shares was carried out, which, even though it was publicized in the local press and through the issuance of brochures, had little success. The political context did not facilitate investment in this type of business and the Civil War paralyzed all the activities that the company was carrying out. However, the almost absolute lack of expenses allowed its corporate assets not to suffer losses, so that although the life of the company was interrupted until 1940, the interests of the shareholders were not diminished.

In the early 1940s, the company reached agreements with the Malaga City Council and its Urban Expansion Commission: in exchange for free land transfers and cash contributions, the road from Torremolinos to the beach would have a favourable location for the company. In 1942, Malaga City Council approved the project for the aforementioned road, which offered them promising prospects since it made it possible to start the construction works of the projected spa. In 1944 the company increased its number of partners and set a new goal: building a hotel. In order to achieve the different social objectives, a series of urbanization works are being carried out, and an important construction activity is carried out, a series of houses for tourist rental are being built. On the other hand, after the sale of land to shareholders, they also build houses, which had to follow the rules established by the social entity to achieve urban harmony. The plans of all the houses to be built had to be approved by the Company Board of Directors and at least two-thirds of the lot had to be left free for the garden. The company guaranteed the right to inspect the works.

In the second half of the 1940s, uncertainty about the buildability of the land slowed down the development of corporate objectives. The Urban Planning Commission of the Malaga City Council began to work on an urban planning plan for Torremolinos, and when the preliminary draft of the same was made public in 1948, almost all of the company's lands were converted into green areas, which completely annulled the corporate purpose. Finally, after the approval of the aforementioned plan in July 1950, although some lands destined for green areas were released, those bordering the beach were absorbed for the layout of the promenade, and therefore unused to undertake the initial social purpose, construction and operation of a spa on those beaches. This led *Torremolinos, S.A.*, to a delicate social and economic situation and to its practical paralysis, since the only way of income would be the sale of land, something that in the long term would lead to the decapitalization of the company. This stoppage continued until 1955, when its dissolution was approved, thus ending the life of a company that was born from the initiative of some people who knew how to see Torremolinos tourist potential in advance.

6 Conclusions

The importance reached by the Costa del Sol in the global sphere of tourism has been the result of the main asset of the province of Malaga, its climate. This, together with the good work of a business fabric that knew how to be ahead of its time, has been the necessary argument to develop a complete tourist offer. During the first half of the 20th century, the Costa del Sol gradually established itself as a tourist destination linked mainly to holiday tourism. This conformation was favoured by the development of communications and the establishment of the first golf course. The airport favoured the arrival of tourists by laying the basis for future mass tourism, and the practice of golf is today one of the great drivers of tourism on the Costa del Sol. Finally, the work carried out by private initiative in relation to the offer of accommodation began to transform the appearance of this part of the Malaga coast, turning it into a large holiday centre. And although some of the projects, which were equal in complexity to those of the main European tourist leisure centres, were frustrated by different circumstances, most of them political, we must give deserved recognition to those pioneers who knew how to identify the climate, the beaches and the sun of this part of the Mediterranean coast, as the great assets of one of the main Spanish industries: tourism.

7 Sources

Archivo Díaz de Escovar (Fundación Unicaja). Málaga

Archivo Histórico Municipal de Antequera, Fondo de Archivos Familiares (family archives)

Archivo Municipal de Málaga
- *British Colony Gazette*
- *El Cronista*
- *La Unión Mercantil*

Biblioteca Virtual de la Prensa Histórica
- *El Defensor de Granada*
- *El Telegrama del Rif*

Bibliothèque Nationale de France
- *Le Petit Marocain*

Biblioteca Virtual de la Provincia de Málaga

Índice Ilustrado de Excursiones. Sociedad Excursionista de Málaga Boletín Oficial del Estado

Hemeroteca de ABC
- *ABC*
- *Blanco y Negro*

Hemeroteca de la Biblioteca Nacional de España
Mundo Gráfico
Nuevo Mundo
El Sol

References

Aguado Pacheco, M. & Utrilla Navarro, L. (2010). *La arquitectura aeroportuaria malagueña. Eugénesis, estética y funcionalidad del nuevo edificio terminal del Aeropuerto de Málaga. Un Aeropuerto para el Futuro*. Madrid: Plan Málaga.

Arenas, Andrés & Majada, Jesús (2003). *Viajeros y turistas en la Costa del Sol*. Málaga: Editorial Miramar.

Boyer, Marc (2002). "El turismo en Europa, de la Edad Moderna al siglo XX." *Historia Contemporánea* (25), 3–31.

Burgos Madroñero, M. (1974). "Málaga siglos XVIII-XIX: Los extranjeros." *Jábega* (7), 49–52.

Koethke, Federico (1929). *Guía del turista. Andalucía*. Málaga.

Luque Aranda, M. (2015). *El desarrollo del sector turístico durante la Segunda República y el Primer Franquismo: La Federación Española de Sindicatos de Iniciativa y Turismo*. Málaga. https://riuma.uma.es/xmlui/handle/10630/10064

Utrilla Navarro, L. (1994). *Historia del Aeropuerto de Málaga*. Málaga: Aena.

Utrilla Navarro, L. (1998). "El Aeropuerto de El Rompedizo: Primera gran puerta abierta al turismo del siglo XX." In VVAA, *Historia de la Costa del Sol* (pp. 49–56). Málaga: Diario Sur.

Zarca, A. (1998). "Del Club de Campo a la Costa del Golf." In VVAA, *Historia de la Costa del Sol* (pp. 81–88). Málaga: Diario Sur.

Margarita Dritsas

"White Flowers" of the Aegean. Would Le Corbusier use the same expression today?

Abstract The chapter focuses on the history of early and later development of Greek coastal tourism in its main expressions: A. coasts and beaches on the mainland and B. the kaleidoscopic clusters of islands in the Aegean sea and the beginning of cruise tourism. The antecedents and impact of early policies introduced by the State – appropriate legislation, introduction of a new culture, availability of human resources, communication – will be assessed. Its nature seems to have been influenced by its conditions of birth: Feelings of national liberation and social freedom, national emergencies, inadequate financial resources, eager entrepreneurship have initially led to a welcome small-scale development, which has soon been followed by global trends contributing to a non-stop process of growth based on high quantitative returns on the one hand and often on rather mixed qualitative results, on the other. Initial progress depended on necessity and the realization that ample natural resources were available for quick development and at low cost with scarce finance. Funds eventually became available as part of the Marshall Plan which gave an impetus to the national effort for recovery of a deeply wounded country/society in the aftermath of WWII and a vicious Civil War. New institutions founded parallel to direct State action had an important part to play in designing and financing the new initiatives suggested by foreign advisors and promoted with foreign investment and lending. Beach tourism, island hopping, coastal shipping, new types of accommodation, secondary middle class holiday residences, new mentalities were some of the main features of the new initiative. Coastal tourism gained momentum, it became a dominant form of travel and vacation long before the 20th century expired. Its central position as Mediterranean travel and tourism raises important questions about environmental sustainability, and whether far from glorifying democracy it rather creates one more divide between mass and elite tourism....

1 Introduction

Greece has been a destination for European travellers for over three centuries. Grand tourists admired or were curious both about nature and classical history. They were also mobilized by the Greek struggle for Independence and the birth of the Greek nation-state. The Grand Tour faded out as travel became easier thanks to the industrial revolution, the number of travellers to Greece grew along with philhellenism until after WWI when the first institutions for tourism were established. Dritsas, M. (2003); (2009); (2016).

Coastal Tourism in the form of beach development, travel to islands and cruises which make up today the core of the sector has received little attention with regard to its historical antecedents in Greece. It is generally taken for granted that it started in the 1960s (Logothetis (1982); Loukissas (1982); Tsartas (1992), (1998), (2003), (2010); Apostolopoulos (1999).

In what follows, this hypothesis is challenged and it is argued that coastal tourism was planned earlier, that it was less connected with domestic tourism which grew much later. Its full growth was the result of a joint rescue plan for Greece at the end of WWII negotiated by USA and Greek authorities. Enforcement of the plan depended on the Greek bureaucracy, special institutions, and rather belated spontaneous, unregulated response from the wider society which shared a particular perception about foreign tourism.

Greek Coastal tourism took various forms and had different rhythms depending primarily on geography, historical circumstances and human agency. Beach tourism, island and sea tourism had different timings and evolution processes as will be seen in what follows.

Coastal tourism developed both on the extensive coasts of mainland Greece and on the plethora of islands in the Aegean Archipelago and the Ionian Sea totaling in length over 16 000kms. Beach development along mainland coasts and islands as tourism destinations for incoming foreign tourists on the one hand and for domestic tourism on the other, had different timing and differential effects. The paper focuses A. On the development of beach tourism in three areas, Chalkidiki in Macedonia, Northern Crete and SW Attica B. On early tourist development of three Aegean islands underlining their diversity and identity as important factors for the growth of various kinds of tourism, including C. The early introduction of sea cruises in the Greek Archipelago.

2 Antecedents of coastal tourism

Towards the end of the 19th century. Thomas Cook introduced organized sea voyages to the Mediterranean (Egypt and the Holy Lands) which also included Greece, while travel guidebooks included information about sailing to the Aegean islands. Travel experiences to and from the East Mediterranean in smaller boats and visits to the islands for leisure are also found in many travel narratives at the end of the 19th century (Buchon, 1841; Sanford, 1907; Tomkinson, 2002; Manatt, 2015; Dritsas, 2019).

By then, and within the climate of political divisions on the one hand and modernization ending in national default on the other, Greek tourism acquired some of its basic characteristics and functions. Coastal/sea leisure pursuits unlike further north in Europe did not concern the Greek population in general, certainly not the farmers, or fishermen or the still weak working class of a predominantly rural country and economy in Europe (Furlough, 1998; Walton, 2011, 2014). Foreign travellers were appreciated by the authorities as contributors to revenues allowing the maintenance of a healthy balance of payments on the one hand and on the other, as messengers praising the progress and virtues of the new nation, or to use a modern analogy, as influencers of foreign public opinion in Europe and beyond regarding Greek national interests, which were at danger very often. This dual function of foreign/incoming tourism was sustained during the interwar and had an impact on the decisions about the reconstruction of Greece after WWII planned jointly by American and Greek advisors.

A similar perception about tourism was also reflected during the interwar years with regard to medicinal/mineral water tourism. Certain locations had become popular and were used without much difficulty by locals. It was however, affluent members of the Greek Diaspora spending their long holidays in Greece who became regular users and raised their popularity and their number. Already the state-founded Greek Tourism Organization in 1929 had a plan for modernization of old and opening of new centres and locations. Towards the middle or end of the 1930s the Metaxas semi-fascist regime trying to gain popularity promoted the idea of free use of spa centres by employees of banks and other important business firms as a sort of bonus in the form of health tourism. After the war new spa centres were founded and older ones were modernized by the new state-run Greek National Tourism Organisation (GNTO). In coastal towns like Aedipsos, Methana, Loutraki or islands (Kythnos, Ikaria, Lesvos) where mineral water springs had been developed, hybrid forms of spa and sea tourism emerged. Dritsas, M. (2002); Hellinikos Organismos Tourismou, Diefthinsis Iamatikon Pigon (1954); Ministry of National Economy. Department of Foreigners and Exhibitions (1930).

Coastal tourism was specifically and deliberately planned and designed after the end of World War II, while the civil war was still raging. Reconstruction needs on the one hand and political priorities by both foreign (USA) and Greek advisors were determining factors. Greece was eligible for foreign aid within the Marshall Plan for the Reconstruction of Europe established in 1947.The extent of damage between 1940 and 1948, on the one hand, and political priorities determined by the Cold War climate already established on the other, were taken into consideration for the allocation of funds. For the first time tourism was recognized as the third most important autonomous economic sector. Development, however, had to be reasonable; as a result, low-cost solutions were included in the Plan favouring the intensive exploitation of what was for the first time considered inexhaustible natural resources like the sun and the sea, demanding, according to foreign advisors, limited amounts of capital. From the beginning, therefore, sea and coastal tourism was seen as a low-cost development/investment process which would attract large numbers of tourists and secure high enough state revenues. The plan was drawn with the participation of Greek experts already involved in tourism earlier who proposed improvement of infrastructure in general, new hospitality units, improvement of archaeological sites, redress of coasting shipping which was totally destroyed during the war, and further development of Rhodes (where tourism had grown during the Italian Rule of the Dodecanese (1912–1945). (Logothetis, 1961).

Despite the many lacunae in the conception of the programme and the ignorance of culture and structure of Greece by foreign advisors, despite inadequacies of the Greek state bureaucracy, the deficiency of political personnel at the time and world political priorities set by the Truman Doctrine related to the Cold War, the programme helped the the Greek economy by assisting industry and promoting tourism, especially after the monetary stabilization of 1954 (devaluation of the drachma) which made Greece a very cheap country to visit. A total of two billion USA dollars were spent on Greece in general modernizing and renewing infrastructure and utility networks destroyed during the War, reviving trade, renewing supplies and restoring shipping, without which island development and growth – included in the 1948–52 plan – would have been impossible. Growth continued until the mid-1970s and the first oil crisis.

Unlike in the rest of Europe, in Greece domestic tourism grew moderately. Furlough, E. (1998). During the period of the military dictatorship (1967–1974) lack of expertise, widespread opportunism, indiscriminate lending policies and corrupt practices caused serious damage. The end result was widespread anarchy and disrespect as regards effective regulation about land use practices and legislation, basic infrastructure requirements, financial rigor, squatting, respect of legality with regard to secondary residences, quality of hospitality facilities, transport improvement and other services.

Domestic tourism started growing after the return to democracy when a more coherent set of social reforms was finally introduced, among them the institution of annual paid leave and health services (Kakoudakis & Papadoulaki (2021). However, the rather populist one-sided policies of the PASOK government did not help in the long run the modernization of the economy or building a healthy financial profile and only increased the state deficit. Finally, the integration of Greece in Europe as a full member in 1981 gave a further boost to foreign tourism, with many Europeans travelling and/or buying property on the mainland and the Greek islands. On the other hand, European policies strengthened de-industrialization which produced negative effects on expertise and employment and led to a different system of monoculture, i.e. serious dependence of the Greek economy on tourism and lack of versatility.

3 Beach tourism

Specific policies after the end of WWII determined the development and opening of organized beaches along coasts in mainland Greece and on Crete. Each area had different social and economic characteristics and attracted different categories or groups of travellers.

Early initiatives came from outside Greece, a notable example being the Club Mediterannée founded by French entrepreneurs with limited finances on the islands of Corfu and Euboea hosting young foreign tourists already in the late 1950s. Furlough, E. (2009). In Crete, it was also mainly foreign tourists who started to travel by air from the mid-1960s on trips organized by global tour operators. In Chalkidiki (Northern Greece) on the other hand, it was domestic tourism especially residents of Thessaloniki – the second largest city of Greece – and progressively from other towns of Macedonia and Thrace who visited the area for short breaks or longer stay. More coastal areas of Chalkidiki became popular as weekend destinations or longer holidays. They were easily accessible by car or bus and affordable for the growing urban middle strata. Accommodation in the area was inadequate. A *Xenia* motel was built as late as in 1962 and a modern camping opened in Paliouri, a village and a community in Kassandra (the westernmost promontory of Chalkidiki) by the Greek National Tourism Organisation (GNTO) founded in 1951. The motel was one of several *Xenia* hotels and guest houses built around the country in the 1950s and 1960s forming a modern hospitality network. Planning and construction was GNTO's responsibility as was for several years also management. When this arrangement proved inefficient, those units were eventually leased out on long leases, to private entrepreneurs, often GNTO's staff or engineers involved in the past in their construction.

Until the 1970s, Chalkidiki had been visited for holidays mainly by residents of Thessaloniki and other towns of Macedonia and Thessaly, many of whom had acquired property there and built summer residences. Thessaloniki had for centuries maintained its international character which was lost after the Balkan Wars. In 1925 an effort to regain it was the new institution of an annual International Trade Exhibition which reunited the city with the rest of Europe but was interrupted during the War and Occupation years. A vigorous campaign by GNTO and local tourism entrepreneurs after the War targeted tourists from the USA and Western European markets, but it was the democratization of

Eastern Europe that reunited Thessaloniki with the Balkan countries and the rest of the world as the gate to Chalkidiki's sandy beaches and wild nature. During the 1970s several camping sites were organized in Chalkidiki and other areas like Pieria South of Thessaloniki towards Thessaly. GNTO organized the first large scale camping site south of Thessaloniki in 1962, where already a beach had been developed. Other camping sites in Chalkidiki were established in 1970 in Paliouri, in 1973 in Kryopigi and in following years near other major Macedonian towns like Kavalla and in Alexandroupoli (Thrace). There were also other simpler camps which developed in the late 1970s.

Another reason for Greeks and foreigners to travel to Chalkidiki, was a pilgrimage of Greek Orthodox and other Christians to its easternmost promontory of Mount Athos (or Holy Mountain), site of a semi-autonomous monastic community of Greek Orthodox monks started in 963 and inhabiting today 20 monasteries and dependencies (*skites*). Visits to Mount Athos have over the years grown in popularity and monks in several monasteries allow visitors to stay for a few days within the community. One such visit was catalytic for the future of tourism in Chalkidiki and its upgrading to international destination.

In 1963 the shipowner John Carras was travelling by sea in the North Aegean aboard his yacht on the way to Mount Athos. The occasion was the celebration of the millennium of the Byzantine monastic community. On that journey Carras sailed towards the Sithonia promontory (one of three such formations in Chalkidiki) known for its beautiful landscapes and wild nature). He anchored in a bay and started planning his next project. Soon afterwards he negotiated the sale of the land with the owner, Gregoriou Monastery. In time, an original self-sustained resort was set up named "Porto Carras." By the 1970s, Carras's next project became building a hotel on the land, the masterplan of which was drawn by Walter Gropius. Priority was also given to the construction of a marina, suitable for large yachts, the first in northern Greece. In time, more innovations were introduced so that the resort and hotel would be adequately equipped to host guests and organize international events and congresses. In June 2003 Porto Carras hosted the European Union Summit Meeting. Resort hotels started to proliferate since in Chalkidiki and elsewhere, becoming a central feature of up-market tourism and "all-inclusive" hospitality models.

Chalkidiki being easily accessible by car acquired high popularity very quickly. A new boom occurred from the late 1980s, the liberalization of Eastern Europe and EU's enlargement playing an important role. Today, tourists from many parts of the World choose the north Aegean sea and Chalkidiki for their long summer holidays and invest in the area.

John Carras was not the only shipowner who invested in tourism in those early days or later. Other Greek shipowners acted as investors in the Aegean islands particularly in cruise shipping.

4 Beach tourism in Crete

Compared with Chalkidiki, Crete has a completely different story. Large scale beach development was organized on the north coast of Crete from the beginning as part of a mass tourism plan. Many kms. of sandy beaches became a lure for low-cost airborne tourism, actively promoted by global tour operators who had and still have a catalytic role on transport, planning and management of a substantial part of the Cretan hospitality industry. Between 1965 and 1973 several projects contributed to the consolidation of Crete's transformation. Tourism infrastructure for the island was for the first time included in the Greek National Tourism Organisation's (GNTO) plan founded in 1951. It included several hotels built in the three main towns on the island's north coast: Chania, Rethymno, Herakleion, and hostels in inland locations (Katsigiannis, 2017). Road construction followed, lighting of major roads and parking lots were provided for the major archaeological sites of Knossos and Festos. GNTO information bureaus opened in the capital Chania and in Herakleion as well as tourism centres of information, and other kiosks in archaeological sites. A School for tourism studies was also established in Herakleion. Other hospitality works planned in the early 1970s like the construction of several hotels west of Chania of a gigantic 1 862 bed capacity, a camping for 500 vehicles nearby, an organized beach and a marina were never completed. GNTO responsible for the project, started from the road construction, water and sewage systems which were also completed before the plan was shelved in 1974 after the collapse of the dictatorship, and in the wake of changes in GNTO's top directorate. For many years, strong criticism had been raised against mass tourism, which although has produced strong critics, seems to have grown roots in Crete as well as in the life and expectations of Cretans.

Against mass tourism, there have been alternative approaches and private initiatives. In the late 1960s, just at the moment tourism was starting to invade the island, the young architect Spyros Kokotos and his wife Eliana, (daughter of a hotelier) decided to buy land, while it was still affordable, in Elounda, an undeveloped area formed of small villages, an uninhabited peninsula and the also uninhabited Spinalonga island (a lepper colony between 1903 and 1957). They continued to buy land in the area even after Kokotos built his first modern luxury hotel "Elounda Mare" in the early 1980s, following the model of bungalows, each with an individual pool and offering quality hospitality. In 1992 the second hotel "Porto Elounda Golf and Spa Resort" opened its doors. It included the first golf course on Crete and introduced time-sharing of partial ownership for seafront villas with private pools. In time, more services and annexes were added including a Spa "Thermais." In 2002 the "All Suite" Hotel upgraded services, offered a cinema theatre, a wine cellar from the Kokotos vineyards, a gourmet restaurant with home grown products and other facilities including the Aegean Conference Centre which contributed to expanding considerably the tourism season by hosting world academic and scientific events.

5 Beach development in Attica/Athens

In the Attica region which concentrates today almost half of Greece's population, and where Greater Athens keeps expanding, leisure and tourism business started long before WWII and became the prelude for the advent of organized leisure on coastal areas along the Saronic gulf to the SW of Athens. Early coastal development became visible when the modern coastal suburb of Faleron emerged, at the beginning of the 20th century, when Greeks from the Diaspora (from Alexandria, other towns in Egypt and other overseas Greek communities) started to buy land along the coast to build villas for their long summer holidays in Greece. Very soon modern luxury and more modest family hotels were also opened. An early initiative was that by Ioannis Pesmazoglou, a Greek banker from Alexandria interested in large modern projects, who bought land in Faleron near the seashore on which the impressive hotel "Aktaion" was constructed in 1903. It quickly became a landmark for the Athenian elite, competing with "Grande Bretagne" hotel on Syntagma Square in the centre of Athens. Aktaion had 160 rooms fully equipped with electricity, private bathrooms with hot water and a "marina" for leisure boats and small yachts. Pesmazoglou leased it for twenty years to A. Mellier an experienced manager, of the famous "Pera Palace" hotel in Constantinople. Faleron continued to grow and became popular for many innovations, among which its "bains mixes" facilities on the beach, its music kiosk, paved paths for walking and cycling along the seashore. Expensive restaurants opened soon and many more hotels and villas were built along the coast. In the 1960s, as a result of new laws about building, modern apartments and new hotels were built along the coast. Faleron is today a town with a separate municipality but with much less of its old middle-class suburb charm. In the 1970s, a modern beach was organized on the coast expanding further along the Saronic gulf.

Ten kms further south from Faleron, lies Vouliagmeni, today a luxury tourist resort. There the Greek Church owned a lake renowned for the curative qualities of its waters. In the 1930s small rental units were built to accommodate visitors who bathed in the lake, while restaurants and cafes had opened and there was already interest shown by prospective investors for hotel construction and further development. The War, however, put on hold such plans. After liberation the opening of a large public beach in Vouliagmeni for use by the urban population of Athens was a project included in the post WWII Reconstruction plan for Greece (as part of the Marshall Plan for the Reconstruction of Europe). It was consigned and inaugurated in 1960 and in one year revenue from entrance fees had grown by 41 %. Eight years later its capacity was extended with more cabins installed, while more beaches were opened in Attica and the rest of Greece. By 1982, there were 19 coastal areas with organized beaches managed by GNTO, when its executives thought it best to transfer management and service to local authorities. As early as in 1954, not far from the Vouliagmeni public GNTO beach another more ambitious project was incubated. after the Monetary Stabilisation of Greece in the same year, the newly founded by the National Bank of Greece company, *ASTIR Tourism and Hotel S.A.* started planning the construction of a hotel. Four years later in 1958 Greek Parliament approved the development, commissioning the plan to a team of well-known Greek architects. The project included an exclusive beach near the Luxury hotel to be called "Astir Palace." The new Vouliagmeni/Lemos beach (or plage) was inaugurated in August 1960, whereas the first 76 Astir bungalows opened in 1961 and in 1966 the first water-ski centre, a real novelty for Greece at the time, also started operating in Vouliagmeni whereas another eight bungalows were added in 1969 to Astir Palace. The project was financed by the National Bank of Greece which had just then merged with its major competitor till then, with regard to financing large projects for many decades, the Bank of Athens. It was that same Bank under the management of Ioannis Pesmazoglou from Alexandria (Egypt) which over 50 years earlier had financed the construction of the modern Aktaion hotel on the coast of Faleron.

In the early 1970s another innovation was added not far from Vouliagmeni. It was the Glyfada Golf Course constructed on an area of 530 acres leased to GNTO by the Municipality of Glyfada. The area was covered with pine trees, dunes and low and dense wooded vegetation. The project was planned earlier taking into consideration initiatives in other Mediterranean countries, the already existing golf course of Rhodes, constructed by the Italians in 1928, and the smaller sand-course of Agios Kosmas (not far from Glyfada) opened in 1931 and closed down in 1940 (Katsigiannis, 2017). It was, however, the Astir Palace hotel which changed the face and identity of Vouliagmeni into that of a luxury resort with expensive villas and exclusive clubs. Together with the nearby coastal areas of

Glyfada and Kavouri they form today the *"Athenian Riviera."* Astir Palace closed down when it was sold a few years ago by the National Bank of Greece to a foreign fund. It has been thoroughly refurbished and operating under its new name: *Four Seasons Astir Palace Hotel.*

6 Early Island tourism in the Aegean

Just like mainland coasts and beaches, island development was included in the post-World War II Economic Reconstruction Plan, although only Rhodes was specifically mentioned. Reconstruction took longer to function coasting shipping having been destroyed during the war, the economy of the islands having collapsed and many islanders having emigrated.

Greek islands are numerous, have many similarities but also important differences arising from their geography, history and local culture. They belong to different clusters, e.g. North East Aegean islands (Lesvos, Lemnos, Thassos, Samothraki); the Dodecanese (e.g. Rhodes, Kos, Patmos, Symi etc.); Eastern Aegean isles e.g. Samos and Chios); the Cyclades (a complex of 25 islands); the larger islands of Crete and Euboea (Andriotis, 2001) and the Ionian islands (Sanford, 1907). Although bibliography is abundant, it does not always cover fast changes.

This paper focuses on three Aegean islands, Santorini, Mykonos and Andros belonging to the Cyclades cluster. All three were well developed before the War, each in different sectors and activities. They also had specific characteristics and traditions. The question asked is how far their earlier structure, their culture and orientation determined their subsequent identity and performance. Islands have their own identity and history and their population reacts differently to plans and proposals formulated outside their societies.

Certain features are common to most Aegean islands. They had an important shipping and maritime tradition for several centuries. Under Venetian rule between the 13th and 15th centuries they had freedom to sail and trade in the Mediterranean; under the Ottomans they had special privileges. Despite severe losses during the War of Independence their fleets continued to sail and trade throughout the 19th and early 20th century. They also took part in all subsequent struggles for territorial consolidation of the Greek nation. Apart from these common features, each island was different in terms of natural characteristics, geographic position, population structure and society. During WWII they were all occupied by the Italian and German forces and their economy suffered as did their population.

In this paper three out of the 25 islands in the Cyclades complex, Santorini, Mykonos and Andros are examined. They are different with regard to geography, size, constitution of natural environment and type of pre-WWII economy; two of them are today hosts of mass incoming tourism while Andros seems to have followed a different agenda.

6.1 Santorini

Today at the top of world destinations, it was literally by accident that a total makeover in its economy became necessary. Under Venetian Rule since 1204 and under the Ottomans since 1579, the island like most other Aegean islands maintained its autonomy with regard to sailing, trading, local government and religious tolerance. It prospered, despite a volcanic eruption in 1452 which ruined half of the isle. It gained independence in the 19th c., along with the rest of Greece and at that time, it had one of the most powerful naval fleets of the nation. Prosperity grew considerably in the last quarter of the 19th century and continued to grow in the next century. Affluence was reflected among other things on impressive residences often resembling little castles.

During the interwar period, tourism began to be perceived in Greece as a potential source of revenue for the Greek state in the aftermath of WWI and the refugee crisis as a result of the Asia Minor Disaster. By the end of the decade, as the world economy collapsed, interest grew and in 1929 the Greek Tourism Organisation (EOT) was founded with the purpose to promote Greek tourism abroad, especially where there were Greek Diaspora communities in Europe and the USA and plan an appropriate infrastructure and hospitality network at home. When a few years later in 1932, the Greek state defaulted on foreign payments and the economy and politics entered a new period of uncertainty, EOT's programme, although reduced, still supported tourism and hospitality initiatives. It was then that in Santorini, the Syrigos merchant family decided to build a hotel on the site of an old residence they owned near the capital *Fyra*, which they named *Hotel Vulcan* after the volcano, an impressive landmark of the island (Acropolis (Athenian Daily newspaper) 13.5.1935). This initiative could only mean that in addition to their other business, they were investing in tourism too. They certainly must have known that cruises had already started in the Aegean and foreign tourists were disembarking on islands, or sailed certainly towards Delos and Mykonos. On the other hand, the hotel may have also housed businessmen, trading with shipowners or with local merchants like Syrigos or other entrepreneurs, journalists, artists, already travelling to Greece and sailing in the Aegean. Political events, however, changed the situation further when in 1935 the authoritarian regime of I. Metaxas took power. EOT was suspended and a Ministry of tourism was founded instead in 1936, attached to the dictator's cabinet. Nothing more was since heard or written in a national newspaper about the new hotel in Santorini.

During WWII the Cyclades were occupied by the Italian forces in 1941, and by the Germans soon afterwards. A joined garrison was established on the island and military officers probably stayed at the *Vulcan Hotel*. In 1943, the joint garrison building of Santorini was raided by the British Special Boat Service (Squadron) Commandos, active since 1942 in the Aegean (Cyclades and Dodecanese) after which heavy reprisals followed by the German army and many Santorinians were executed. Others found ways to escape or leave the island altogether.

With the return to peace rather later in Greece than in the rest of Europe because of the Greek Civil War, the situation on the island remained difficult. The country on the whole had been devastated, its infrastructure destroyed and Reconstruction could only be undertaken with help from outside (Marshall Aid funds as already mentioned ealier) which became available through the re-instituted GNTO and Greek banks at the end of the decade. In 1950, however, the volcano erupted causing severe damage and five years later, in 1956 a catastrophic earthquake completed the destruction. Many Santorinians left, the island's economy collapsed, others returned years later when regular transport resumed in the Aegean and tourism started with assistance from the Greek state and low interest loans from banks. Later still the archaeological excavations in Akrotiri brought to light unique findings which added value to the profile of the island luring more visitors. Because of the island's land formation accessibility problems persisted, until a new road, a port and an airport were built in 1972, contributing to a quick rise of foreign tourism. New opportunities arose for private initiative especially with regard to the growth of the building industry often leading to unlawful actions, especially after the introduction of legislation for the preservation of traditional villages after the earthquake and later for protection of the shores. Congestion remains a major problem for most islands in the Aegean during the summer months with regard to regular tourists and cruise boats.

6.2 Mykonos

Problems are not very different in another rather small Aegean island which however, monopolizes publicity hosting every year millions of tourists, many of them international celebrities. Under Venetian rule after 1204 Mykonos was conquered by the Ottomans in 1537 and like most other Aegean islands continued to prosper as a trading centre until the 18th century. Its fleet took part in the Greek struggle for Independence against the Ottoman Empire with the constant support from Manto Mavrogenous, daughter of a wealthy Diaspora Greek merchant from Trieste. She equipped her own ships and commissioned others to fight piracy as well as the Ottoman fleet in the Cyclades, in Samos and in Chios.

Mykonos's importance in trade faded when at the end of the 19th century sea routes changed after the opening of the Corinth canal. Many Mykoniates had to emigrate. A new role, however, loomed up for the island when the archaeological excavations in Delos got under way at the turn of the century by the French Archaeology School of Athens (French School at Athens 2017). Passage to Delos was possible only from Mykonos in small boats. By the interwar, publicity, curiosity and scientific interest for Delos transformed the island to a gate leading straight to ancient Aegean culture. At the beginning, visitors stayed in the homes of locals or those of members of the excavation staff, employees and workers. Numerous new houses were also constructed, and with time more sophisticated guest houses and hotels went up. Local manpower was used and many of the buildings respected local vernacular architecture. It was as late as in 1953 within the Reconstruction plan, that GNTO inaugurated its post WWII hospitality project for Greece by building the first hotel, or rather modern guest house, named *Leto* with a capacity of 44 beds. In 1948 an older hotel named *Apollon* received a loan allocated by the new Organisation for Hotel Credit and issued by the National Bank of Greece. Later, Mykonos started to attract art lovers, artists, collectors, socialites and many investors in tourism hospitality and related trades. Locals, Athenians and foreign entrepreneurs invested and in time infrastructure was improved. (Stott, 1973) Then yachts and cruise ships started to proliferate, air transport and fast boats followed suit and Mykonos was/is listed today with the most visited island destinations in the world.

Underneath the shining side, there are nevertheless many problems, common in most islands. They range from bureaucratic feuds, clientelist politics, complications concerning building laws and regulations, confusion about jurisdiction of organizations involved in the running of local affairs and businesses, safety and security issues. Law transgression cases are as frequent as property issues often leading to protracted disputes and court cases, obstructing preventive action also with regard to conservation issues and sustainability practices.

6.3 Andros

The second largest island in the Cyclades, has been chosen because of its very different trajectory and profile compared with Mykonos, Santorini and many of the smaller islands in the complex. Its different natural environment, rich heritage, and strong shipping tradition and culture has interested historians, intellectuals, poets and artists. Shipping contributed to the wealth of the island, has nurtured important personalities and daring innovators, has shaped the identity of Andriots based on pride for the island's history, respect for culture, tradition and heritage and lately on the support of alternative tourism and environmentally friendly initiatives.

Andros like other Cycladic islands impressed visitors since the 17th century. Numerous foreign and Greek travellers left detailed descriptions and comments about its history, its strategic position in the Aegean since the Byzantine times, its powerful fleet and skilful mariners, effective diplomacy practiced and a strong economy based on local production, and last but not least on shipping (Dritsas & Papadoulaki, 2019).

Much later, diplomats, scholars and artists appreciated the beauty and special atmosphere of the island, explored and wrote about it and even painted some of the island's landscapes, e.g. Arthur Tower, an Officer in the English Navy in the 1840s. (Kaïreios Library-Andros 1987).

From the end of the 19th c., the number of visitors grew, and those who wished to stay for longer periods rented houses from the locals usually in or near the quiet, friendly fishing village of Batsi, on the west coast (Manatt, 2015). Batsi gradually grew and is today the main tourist hub of the island. A large hotel (Perrakis), perhaps the older "modern" establishment in Andros overlooks the beach with many smaller and larger units around covering today the wider area of the bay of Batsi.

Andros until the late 1960s was not visited or praised for its beaches or coasts but for its woods, history, culture and rural beauty. The first location which was officially described as a tourist destination on the island in the 1930s was the small town of Sariza, famous for its waters and beautiful nature. Visitors during the interwar years and early post WWII came to the island and Sariza from many parts of Greece, but especially from Attica and Athens to drink the water, just as others visited Loutraki near Corinth. Following the example of Loutraki, in Sariza a bottling factory was established trading the famous water. Guests stayed in local houses for a few days, drank the water according to their medical prescriptions and then returned home. Later, a hotel was built in the area. Sariza remains popular still today because of the beautiful green surroundings, local produce, bird singing and the famous water spring. Visitors can enjoy walking on the paths leading to the spring and beyond that to fields with their preserved dry stone tier enclosures.

Foreign tourists became noticeable in the 1970s, visitors coming mainly from UK and from Scandinavia. The trend is still strong, more diversified as to age, style and origin. Modern entrepreneurs in Andros are now trying to develop new infrastructures enlarging the port of Gavrion and creating new spots for harbouring super yachts, hoping to upgrade other towns too apart from Batsi and Hora.

The island has today over 25 large hotels (over 80 beds each), almost 300 smaller premises (40 beds each) and short-let houses and villas capable to house over 3000 visitors. Before the Covid-19 crisis, walking holidays were chosen by approximately 2000 visitors. In 2018, a 19th century path from Hora to Korthi was rebuilt and is used for walks, whilst another path on the flank of a gorge where a river flows was also rebuilt several years ago with EU funds for conservation works.

The main town, Hora, is family friendly and still preserves its natural environment. For a long time, its residents opposed tourism development but as the situation in the merchant navy began to change fewer positions were/are available to Andriots in shipping. The old tradition that shipowners enlist people from their islands and home towns in their ships is still observed as far as officers but not for lower crew positions any more. The town still has only one narrow main street, old traditional buildings and several museums which preserve much of its old conservatism. It has a permanent population of high calibre, the well-stocked Kaireios library for historical material related to Andros and the Cyclades, and family archives of intellectuals and prominent families of the island open for locals and visitors. In the last few years, land sale in the area has grown, there are many newcomers and older houses are being converted to short-let accommodation, or change ownership. Many Athenians with properties on the island spend long summers there. Local entrepreneurs and authorities already target new markets for future tourism expansion and lengthening the season. A characteristic feature today with regard to the tourist season is that the hot summer months of July and August are preferred by Greek tourists while September to November are chosen by foreigners. Those with property also visit the island for other holidays like Christmas and Easter.

The island panorama in Greece is very wide and kaleidoscopic in nature. Writing about island tourism the diversity is often lost and this is the problem with all major plans about general development. They tend to ignore specificities and usually destroy the preciousness that goes with them. In this paper, only few basic characteristics are mentioned. There have been many interesting and beautiful quotes by visitors, intellectuals and artists highlighting the particular identity/uniqueness of the islands. Le Corbusier's observation (used as a citation and title for this chapter) is still valid today, although white structures on the islands are sometimes large and not always of the vernacular type. One quote however, stands out; it refers to the timeless essence of the Aegean islands as much more than an ephemeral tourism experience. It was written by the Greek historian Vasilios Sfyroeras (1985), born on a rocky village of Naxos in the Cyclades. For him the islands were/are *"treasure boxes of culture and the soul."* (Sfyroeras, V(†)., A. Avramea, A., S. Asdrachas, S. 1985).

7 The beginning of cruises

Andros's power has traditionally laid in the sea and its shipping community sailing and trading across the Mediterranean and the Black Sea for at least two centuries (Beneki, 2006). Early in the 20th century, Leonidas Embeiricos, (trained in the family's branch office in Braila, Rumania), established with his brother the National Steam Navigation Company of Greece with a tradition in cargo shipping. Embeiricos changed that when they decided to invest instead in transatlantic passenger shipping and more particularly in emigration transfers from Greece and other Mediterranean countries to the USA and Canada. Another Andriot, former master mariner, Dimitrios Moraitis had invested in big liners before Embeiricos but did not survive long. Embeiricos was undoubtedly an astute businessman but probably also sympathetic to the problems of many islands and Greece in general, afflicted by the severe social and economic crisis of the early 20th century. He acquired his first passenger vessel which he named "Patris" in England and had her cabins fit to carry more passengers/migrants. By 1925, he had almost monopolized emigration transfers. Embeiricos also got involved in politics and in 1919 he employed Iraklis Ioannidis, a Greek from Asia Minor, as secretary (École Française d'Athènes, 2006). After the fall of the government in 1920, having noted the dynamism and talent of Ioannides proposed that he should undertake the management of his offices in Paris and so Ioannides was put in charge of the Neptos Co. which represented the steamship lines from New York, Marseilles and Piraeus with the company's famous *Patris* vessel (Basch & Farnoux (Dir.), 2006). It was then that the first luxury cruises were organized by Ioannides who promoted them in the pages of a newly constituted periodical entitled *Voyage en Grèce,* published by him directly from 1934 onwards (Dritsas, M. & Papadoulaki K. 2019): Ioannides working from France was well aware of the new trend of cruise tourism for the middle classes already expanding. (Cerchiello & Vera-Rebollo, 2019). An important moment in Ioannides's initiative was the decision in 1933 of the organizers of the 4th Contemporary Architecture Conference to travel aboard PATRIS from

Marseilles on a cruise which would bring them to Greece. In the course of their journey they visited the Cyclades (Basch & Farnoux, 2006). Taking part in this early Aegean cruise was the French architect Le Corbusier who as he set eyes on the islands from the ship rising in the blinding Greek light called them "the white flowers of the Aegean" and sealed in a way the image of a dream island as a barren rock crowned by beautiful white cubes (typical of the Aegean vernacular architecture).

Embeiricos already experiencing problems, two years later, decided to sell his whole fleet. Ioannides dismayed by the decision decided to continue on his own, by setting up a travel agency named *"Le Voyage en Grèce"* which organized and promoted cruises under the title "Les Escales d' Ulysse," chartering Greek-owned ships. It was the same travel agency which later started collaborating with the first airline flying to Greece under the name "Hellas." Nevertheless, because of the continuing Civil War in Greece, Ioannides never succeeded in launching more innovations or take new initiatives with regard to Greek tourism.

Cruises in the Aegean restarted after 1954 by George Potamianos, second generation shipowner from Kefalonia in the Ionian Sea, taking in a way the lead from Embeiricos and securing GNTO's support in 1953. His ship named Semiramis (*previously known as Calabar*) started round trip cruises to the Greek islands and the Eastern Mediterranean. *She* was built in England and was bought by A. Potamianos's *Epirotiki Lines* in 1953. She was the first Greek-owned cruise vessel sailing in the Aegean after the War, after Potamianos's decision to focus business entirely on the cruise ship market, which was then undergoing a transformation. No longer a privilege of the wealthy, it became a vacation option affordable by a larger, middle-class traveling public. In 1955, cruises organized by GNTO were advertised as *"Croisières aux îles enchentereuses de l' Egée"* par paquebot Sémiramis. In other posters, Greece was called *"Pays des Merveilles, Pays de Légendes."* Other slogans inspired by legend were used by GNTO a few years later, for example, *"Let the Breath of Aeolos Move you."* The initiative was successful and in 1957 the ship *Hermes* also owned by Epirotiki Lines was added to the fleet. She was previously chartered by foreign travel agencies for cruises in the Mediterranean and the Black Sea. In the 1960s, *Epirotiki* began expanding by adding a number of Caribbean destinations, whilst for Greece, most of the company's business was during the months of March through November for cruises in the Eastern Mediterranean and destinations to the Greek islands. Meanwhile, in 1958, *Sun Line Maritime Co.* owned by Charalambos Kiosseoglou launched his *Stella Maris*, also a cruise ship, while Markos Nomicos added his *Delos* for Aegean cruises. In 1958, more vessels were added to the Greek cruise fleet owned by Chandris Shipping Co. and GNTO continued organizing cruises to the Aegean islands focusing mainly on foreign tourists and on beautiful locations and places or islands with archaeological interest (Foustanos, 2008 and 2010).

The change towards socialist governance in 1981 ushered in income redistribution policies and social reforms for the middle and working classes, justifying the decision of those shipowners who launched cruise tourism in Greece. The change in policy resulted in boosting not only incoming foreign tourism but also domestic tourism which had remained less developed after WWII. Cruise tourism grew further and Epirotiki Lines became the largest cruise shipping company in Greece and the Eastern Mediterranean. *Semiramis* continued to sail both as a coasting ship servicing Aegean islands and/or as a cruise ship until it was sold in 1979 to Saudi Arabia making room for larger and more modern transatlantic cruise liners. In 1987, there were 47 Greek owned cruise ships serving 10 344 passengers and employing 1 303 personnel. Tourism Receipts (in million drachmae) from cruises for the period 1987 to 1989 were: 1987=27.005; 1988=37.339; 1989=44.509 representing 5 %, 6.8 % and 7.7 % of total tourism receipts for those three years (Pagonis, private papers 2021.)

Meanwhile yachting had also risen as a new form of leisure and many smaller wooden yachts were also engaged for visits and cruises to the Cyclades. Updating in a way the pre-war French/Hellenic initiative of *Voyage en Grèce* and the fascination of Le Corbusier, Constantinos Doxiades, a world known Greek city-planner in the early 1960s started organizing an annual international event entitled "Delos Floating Symposium." It was an international workshop meeting on modern city-planning taking place aboard a wooden vessel sailing in the Aegean towards the Cyclades.

In the following decades, cruises and yachting prospered by gaining enormous popularity both among foreign and domestic tourists. It is certain, that any journey to the Aegean islands has a particular charm for any traveller. Le Corbusier was not the first to be impressed. In 1879, Joseph Reinach wrote: "the islands are Oceanids dancing around Poseidon... the Cyclades stand out ruggedly like black figures on Greek pottery, light floods me... the soul awakes, I feel triumphant" (Dritsas, 2019a).

8 Conclusion

What preceded is an attempt to give an outline of the early beginnings of three basic forms of coastal/sea tourism, i.e. beach development, island tourism and cruises in the Aegean. Greek mass tourism development has generally been considered by tourism experts and many Greek scholars as starting in the 1960s, that is, long after the Second World War. In what preceded it has been argued that the initial idea was born earlier, based on advice by foreign (USA) and Greek experts within the new at the time coordinates of the Truman Doctrine and the conditions and funds specified and allocated by the Marshall Plan for the Reconstruction of Europe. There was never any clear distinction between incoming foreign tourism, on the one hand, and domestic tourism on the other, but the emphasis was placed on projects which would increase revenues for the Greek State and the economy, especially in foreign currency.

Of the three areas chosen for examination, beach tourism was thought by the planners as development for both foreign and local tourists, although domestic tourists were few, Greece still being largely rural and poor (with the exception of Athens). During the Interwar, if and when Greeks went on holidays, they preferred the few spa locations and hydrotherapy centres. In the three cases (Chalkidiki, Crete, Attica) infrastructure (and operation for several years) was developed by the state run Greek National Tourism Organisation. Their specific character, however, was shaped through different types of local development and private initiative, i.e. resort locations.

In the case of the islands, diversity was very wide. Santorini destroyed by an earthquake was redeveloped "at the right time" (mid-1950s) quicker than other Aegean islands. Mykonos also started to develop much earlier because of proximity to Delos, with local and emigrant power. Andros followed a completely different path based on its history, good soil, long tradition and strong identity of its people. To this day, initiatives are taken about environmentally friendly pursuits and leisure and about the promotion of culture.

Finally, cruises -not for aristocrats only- also started quite early and developed fast, mainly as part of incoming foreign tourism, reviving a long tradition of sailing in the Aegean by Greek and foreign travellers. A considerable number of Greek shipowners were involved, investing in this form of tourism, especially at a time they were trying to recover from the World Depression and later from damages endured during the War.

In the paper the early stages of coastal/sea tourism development in Greece were analysed both before and after WWII. Findings confirm: the strong diversity of tourism among islands and other locations from the early days until much later; the important role of private initiative; the role of the State after WWII based on the decision to use incoming tourism as a means to increase public revenue and contribute to the stabilization of the economy.

References

Acropolis (Athenian Daily newspaper) 13.5.1935.

Andriotis, K. (2001). "Tourism Planning and Development in Crete: Recent Tourism Policies and Their Efficacy." *Journal of Sustainable Tourism* (9–4), 298–316.

Apostolopoulos, Y. (1999). "From Farmers and Shepherds to Shopkeepers and Hoteliers: Constituency-differentiated Experiences of Endogenous Tourism in the Greek Island of Zakynthos." *International Journal of Tourism Research* (1) 413–427.

Basch, S. and A. Farnoux (Dir.), (2006). *Le Voyage en Grèce 1934–1939. Du périodique de tourisme à la revue artistique.* Athènes: École Française d'Athènes.

Battilani, P. (2002). "Rimini and Costa Smeralda: How Social Values Shape Recreational Sites." In S. C. Anderson & B. Tabb B. (Eds.), *Water, Leisure and Culture. European Historical Perspectives* (pp. 209–221). Oxford & New York: Berg.

Battilani, P. (2009). "Rimini. An Original Mix of Italian Style and Foreign Models?." In L. Segreto, C. Manera & M. Pohl (Eds.), *Europe at the seaside. The Economic History of Mass Tourism in the Mediterranean* (pp. 104–124). New York & Oxford: Berghan.

Beneki, E. (2006). "From the Aegean to the world: Merchant and shipping activities of the Embiricos family from Andros, mid 19th to mid 20th century." In M.C. Chatzioannou & G. Harlaftis (Eds.), *Following the Nereids. Sea Routes and Maritime Business, 16th-20th Centuries* (pp. 112–117). Athens: Kerkyra.

Buchon, A. (1841). "Excursions Historiques dans les Cyclades: îles de Tinos et Andros 1841." *Revue Indépendante* 25 avril 1844. Tome XIII. Et Emile-Paul, 1911 (pp. 563–571). Paris: Longnon.

Cerchiello, G. & Vera-Rebollo, J.F. (2019). "From Elitist to Popular Tourism: Leisure Cruises to Spain During the First Third of the Twentieth Century (1900–1936)." *Journal of Tourism History* (11–2), 144–166.

Dritsas, M. (2002). "Water, Culture and Leisure: From Spas to Beach Tourism in Greece during the Nineteenth and Twentieth Centuries." In S. C. Anderson & B. Tabb B. (Eds.), *Water, Leisure and Culture. European Historical Perspectives* (pp. 193–208). Oxford & New York: Berg.

Dritsas, M. (2003). ''Tourism in Greece: a Way to What Sort of Development?" In L. Tissot (Ed), Development of a Tourism Industry in the 19th and 20th centuries. International Perspectives (pp.187-208). Neuchâtel:Editions Alphil.

Dritsas, M. (2009). ''Tourism and Business during the Twentieth Century in Greece: Continuity and Change."In L.Segreto, C. Manera and . Pohl (Eds.),

Europe at the Seaside. The Economic History of Mass Tourism in the Mediterranean (pp.49-71). New York & Oxford: Berghahn Books.

Dritsas, M. & H. Coccossis (Eds.), (2014). Tourism and Crisis in Europe XIX-XXI centuries. Historical, National, Business History Perspectives. Economia Publishing. Athens 2014.

Dritsas, M. (2016). "Outline of Tourism in Greece During the Twentieth Century: Continuity and Change." *Revista de la Historia de la Economía y de la Empresa*, 10, 53–84.

Dritsas, M. (2019a) "Andros sto chrono: Maties periigiton." (Andros Through Time: Through the Eyes of Travellers). in Dritsas, M. & Papadoulaki, K. (2019) *Psifides Historias tou Hellinikou Tourismou*. (*Mosaic of Greek Tourism History.*) Athens: Economia Publishing.

Hellinikos Organismos Tourismou, Diefthinsis Iamatikon Pigon (Greek Tourism Organisation, Mineral Springs Directorate). (1954). *Loutropoleis kai Iamatikai Pigai (Resorts and Mineral Springs)1951-1952-1953*. Athens: National Printing-House.

French School at Athens (2017). *1873–1913 Delos. Images of an Ancient City Revealed through Excavation*. Athens: Melissa.

Furlough, E. (1998). "Making Mass Vacations: Tourism & Consumer Culture in France, 1930s to 1970s." *Comparative Studies in Society and History* (40–2), 247–286.

Furlough, E. (2009). "Club Méditerranée 1950–2002." In L. Segreto, C. Manera & M. Pohl (Eds.), *Europe at the Seaside. The Economic History of Mass Tourism in the Mediterranean* (pp.174–195). New York & Oxford: Berghan.

Foustanos, G. M. (2008). *A Century of Greek Passenger Ships*. Athens: Argo Publishing.

Foustanos, G. M. (2010). *Greek Coastal Service*. Athens: Argo Publishing.

Hundstad, D. "A 'Norwegian Riviera' in the making: the development of coastal tourism and recreation in southern Norway in the interwar period." *Journal of Tourism History* (3–2), 109–128.

Kaïreios Library-Andros. (1987) *Water-colours of Andros (c. 1840): The Tower Album*. Athens: Agra Publications.

Kakoudakis, K. I. & Papadoulaki, A. (2021). "Social Tourism in Greece: A Brief History of Development from the Interwar Years to the Covid-19 Era." In J. Lima & C. Eusébio (Eds.), *Social Tourism: Global Challenges and Approaches* (pp. 5–17). Oxfordshire: CAB International.

Katsigiannis, K. (2017). *Hellinikos Organismos Tourismou Taxidi sto Chrono* (*Greek Tourism Organisation, A Journey through time*). Athens: Private Publication.

Larrinaga, C. (2016). "El impacto económico del turismo receptivo en España en el siglo XX, 1900 a 1975." *Revista de la Historia de la Economía y de la Empresa* (10), 23–50.

Logothetis, M. I. (1961). *O tourismos tis Rodou (Tourism in Rhodes).* Athens: Lavyrinthos.

Loukissas, P. J. (1982). "Tourism's Regional Development Impacts. A Comparative Analysis of the Greek Islands." *Annals of Tourism Research* (9), (523–543).

Manatt, James I. (1914) *Aegean Days.* (In Greek: *Meres Aigaiou, Ena kalokairi stis Kyklades.*) Kifissia: AlphaTrust 2015).

Ministry of National Economy. Department of Foreigners and Exhibitions (1930). *Deltion Iamatikon Pigon* (Mineral Springs Bulletin) (1). Athens: National Printing-House.

Pellejero Martínez, C. (2009). "Tourism on the Costa del Sol." In L. Segreto, C. Manera & M. Pohl (Eds.), *Europe at the seaside. The Economic History of Mass Tourism in the Mediterranean* (pp. 206–232). New York & Oxford: Berghan.

Sanford M. P. (1907). *Greece and the Aegean Islands.* London, Boston & New York: Archibald Constable-Houghton, Miffin.

Sfyroeras, V(†)., Avramea, A. & Asdrachas, S. (1985). *Hartes & Hartografoi tou Aegeou Pelagous (Maps and Map Makers of the Aegean Sea).* Athens: Olkos.

Stott, M. A. (1973) "Economic Transition and the Family in Mykonos." *The Greek Review of Social Research* (17), 122–133.

Tsartas, P. (1992). "Socioeconomic impacts on tourism in two Green isles," *Annals of Tourism Research* (19), 516–533.

Tsartas, P. (1998). *La Grèce: Du tourisme de masse au tourisme alternatif.* Paris: L' Harmattan.

Tsartas, P. (2003). "Tourism Development In Greek Insular And Coastal Areas: Ociocultural Changes And Crucial Policy Issues." *Journal of Sustainable Tourism* (11-2-3), 116–132.

Walton, J. K. (2000). *The British Seaside: Holidays and Resorts in the Twentieth Century.* Manchester: Manchester University Press.

Walton, J. K. (2011). "Seaside tourism in Europe: Business, Urban and Comparative History." *Business History* (53) 6, 900–916.

Walton, J. K. (2014). "A paradox of the Inter-War Depression? Markets and Patterns of Innovation in Coastal Tourism during the 1930s: An International Analysis." In M. Dritsa(s) (Ed.), *Tourism and Crisis in Europe XIX-XXI Centuries* (pp. 20–38). Athens: Economic Publishing.

Petra Kavrečič and Metod Šuligoj

"Socialist-style tourist accommodation"

Abstract The chapter will present the development and assertion of a type of social tourism after World War II in socialist Yugoslavia. This was an important phase of tourism development, especially in the area around the Adriatic. The case study will address the Slovene seaside, where, since the mid-1950s and especially after solving the Yugoslav-Italian-border issue, authorities promoted the establishment of union managed holiday homes in other adapted tourist facilities. A special type of social tourism was oriented towards social groups that were traditionally overlooked – i.e. workers from all parts of Slovenia (Yugoslavia). Various companies and public sector organizations have built or opened holiday homes for their employees, in order to grant them some gratification for their hard work throughout the year. The process of social accessibility of tourist services was granted by the state. The Yugoslav government introduced paid leave in 1946 and "*regres*" (annual holiday allowance) in 1965. The new political regime significantly emphasized the social role of tourism, which also gained ideological connotations: "*Tourism and leisure should belong to working people to gather new strength for further work.*" Practices of the domestic elites were, however, not completely in line with this idea.

1 Introduction

> Our wealth is not just factories and roads.
> Our wealth is a man, a new man, a socialist man
> that needs to be developed.
>
> (Josip Broz Tito, the Yugoslav president)

The Upper Adriatic is the northernmost region of the Mediterranean, where empires and neighbouring nation-states ruled: the Austrian Empire/Austria-Hungary until 1918, the Kingdom of Italy (officially from 1920 to 1943), the socialist Yugoslavia[1] until 1991, and then its legal successors (Ashbrook, 2006, pp. 3–4; Žerjavić, 1993, p. 632)[2]. Consequently, in the 2nd half of the 20th century

---

1 The successor state to the Kingdom of Yugoslavia, existed under various names, including the "Democratic Federation of Yugoslavia" (1943–1945), the "Federal People's Republic of Yugoslavia" (1946–1963) and the "Socialist Federal Republic of Yugoslavia" (1963–1991/92).
2 More can be found in Marcks, Knieling & Vladova (2016), Violante (2009) or Reverdito (2009).

there were still great disputes between Italy and Yugoslavia in determining the state border of its northern part (Istria) (Pirjevec, 2007; Kosmač, 2017; Troha 2018 and 2019). Thus, this area can be understood as one of the hot spots of Europe. However, due to its climate, cultural and historical sights, this area has been internationally recognizable since the time of the Habsburgs, as part of the Austrian Littoral crown land (Kavrečič, 2017; Šuligoj, 2015). The accommodation industry also gradually developed in the most important coastal areas of the Upper Adriatic: Gradež/Grado, Opatija/Abbazia, Brioni islands, Lovran/Laurana and Portorož/Portorose.[3] Both smaller and larger hotels such as the Palace in Portorož and the Kvarner (Hotel Quarnero) in Opatija were practically the first flagships of the modern hotel industry in the entire Eastern Adriatic (not only on its upper part). During the period between the two World Wars, marked by the fascist political doctrine, the health and economic crisis, the Kingdom of Italy did not significantly improve the accommodation industry infrastructure and services. However, during the interwar period, the growth of guests was particularly recorded in Portorož. The tourism industry was supported by the fascist regime for different reasons; especially by the fascist political doctrine of the Italianization of the newly acquired (or finally redeemed) lands (Kavrečič and Radošević, 2017; see also Kavrečič, 2020). Nonetheless, the tourist destinations in the Upper Adriatic areas changed significantly in the post-World War II period, when socialist Yugoslavia initiated the construction of a new kind of accommodation facilities (based on the demand) and infrastructure on its coastline. Special attention was paid to the domestic working class, Yugoslav authorities and companies constructed special accommodations suited for them; this period was an important phase of tourism development in the area.

The case study will focus on the North-Western part of the state (the Eastern side of the Upper Adriatic), which at that time administratively belonged to Slovenia, one of the Yugoslav republics and is still part of today's independent state the Republic of Slovenia. On the Slovenian coast, authorities promoted the construction of workers' holiday homes and other adapted tourist facilities from the mid-1950s onwards.

The aim of the paper is to illuminate the specific Yugoslav model of domestic tourism and related "socialist-style tourist accommodation," which were primarily intended for the working class, a pillar of the then socialist society. The post-war period with the systemic focus on workers' tourism gradually developed into mass domestic tourism across the entire Eastern Adriatic. However, in this paper, special

---

3   When we initially mention a locality, it is written in both Slovene or Croatian and Italian. Afterwards, only the Slovene or Croatian nomination is used.

attention is solely given to accommodation in the Municipality of Piran (where the tourist resort of Portorož is located), which has been the most developed tourist area in Slovenia for more than 100 years. Hence, we focus on the positive side of the Yugoslav undemocratic regime, which was quite totalitarian in the years immediately following World War II and during the socialist revolution (Flere, 2012).

2 The border issue and transition to socialism

After World War I, the territory taken into closer consideration was subject to political negotiations. As Italy was actively involved in the war and on the side "of the winners," the promised territories were assigned to the state (Treaty of Rapallo 1920). The Italian fascist regime (since 1922) thus strongly affected this territory, with the aim for a complete national and cultural Italianization (with severe repression on the Slovene and Croat population).

After World War II, the situation in this region was very complex, since the new Yugoslavian state demanded the correction of the borderline between Yugoslavia and Italy, while the Treaty of Rapallo was considered unjust for Yugoslavia. Since this issue was strongly problematic, long diplomatic discussion took place. The border demarcation line, the so-called "Morgan Line" was set up in this region, known as the Julian March. It was the border between two military administrations in the region: the Yugoslav on the east (zone B), and that of the Allied Military Government on the west (zone A). The border issue was partly solved in 1947 with the Peace Treaty of Paris. The Treaty established the border between Italy and Yugoslavia in the Northern sections of the contended territory, as well as the border between the two states and the Free Territory of Trieste, which was established as a new independent, sovereign State. This territory was divided, similarly as the Julian March, into two administration zones (Zone A, under an Allied Military Government and Zone B under a Yugoslav Military Government). In 1954, with the signing of the London Memorandum, both military governments handed over their mandate to the Governments of Italy and Yugoslavia. The issue was finally solved in 1975 (The Osimo Treaty) (Troha, 2018, 2019).

By the end of World War II, and after all the related tragic events, the region being discussed in this chapter began a new period of drastic political, economic and demographic changes. However, the consequences of war had to be overcome first. The new political regime (socialism), the exodus of the Italians[4] and

---

4 Including political emigrants of Slavic origins – see Bufon (2009, p. 460), Pipan (2007, pp. 226–227, 237), Purini (2012, p. 425) and Šarić (2015). Italians were members and followers of the fascist party and capitalists.

the immigration of Slovenes from other parts of the republic and people from other Yugoslav republics,[5] significantly influenced the development of society and the economy on an individual, micro and macro level (see also Purini (2012), Hrobat Virloget (2015), Hrobat Virloget (2021), and Oblak Moscarda (2016)). The new political regime was established in Yugoslavia, the adjusted soviet style communist political, social and economic regime was introduced. The Yugoslav Communist party abolished the liberal democracy and market economy. All economic activities were nationalized.

3 Yugoslav socialism and tourism

Tourism, during the first years of socialism, took a back seat, as the new state government primarily promoted the industrialization and the reconstruction of infrastructure destroyed during the war. In the first post-war years, the nationalization process further hindered the tourist industry. Arrivals of foreign tourists were rare. The problem was in the lack of tourist infrastructure, modest supply, poor investments etc., in addition to the fact that a visa was required for foreign tourists to enter in Yugoslavia. Due to ideological reasons, tourists from Western countries were also not very welcome, at least in the first post-war years; the role of the Slovenian Tourist Board in the first post-war years was marginalized due to the political agenda. Only in the mid-1950s, the Board "began developing into a civil society organization" and work on tourism development (Repe, 2006, p. 62).

Reforms gradually started after the break of political relations with the Soviet Union in 1948 (Expulsion of Yugoslavia from Cominform in June). Yugoslavia also partially introduced or developed specific forms of consumer society. The partly changed economic agenda also affected the tourism sector. In fact, in the 1960s, tourism (not only for workers but foreign, commercial tourism as well) became one of top priorities of the regime (Repe, 1996, p. 160). The right to annual holiday leave for all the population, acquired after the war, encouraged the development of tourist activities as part of the "modern way of life." In fact, "when Yugoslavia was recreated as a socialist state in 1945, the new leadership defined holidays with pay and recreation for workers as crucial elements of the new state's social project" (Taylor and Grandits, 2010, p. 3). In this process, the coast along the Adriatic Sea was the centre of tourist activity in the

---

5   See also Žerjavić (1993), Nejamšić (2014), Medica (2011, p. 250), Violante (2009) and Bufon (2008).

state. Nonetheless, during the second half of the 20th century, seaside summer tourism became the most popular activity in the world. The Yugoslav coast, along with other Mediterranean destinations, offered sea, sun and sand; the "socialist" Yugoslavian coast was also more affordable from a financial point of view (due to the lower standard of living in the state). Tourism developed along the coast as a basically new economic activity along the southern coasts of the Eastern Adriatic (Dalmatia and Montenegro). On the other hand, we should not forget that the Upper Adriatic coastal towns had a long tradition of tourism development, dating back to the 19th century (Kavrečič, 2017; Šuligoj, 2015).

Soon after the establishment of the new regime in Yugoslavia, important reforms in social life were introduced. The system's political doctrine was in fact concentrated towards the workers. The aim was to emphasize working peoples' efforts to build the new regime; to grant the deserved reward for their work. Different social measures were introduced: the annual two weeks paid leave was introduced in 1946 (with extensions of days in the following decades) and "regres" (annual holiday allowance) in 1965. The new political regime significantly emphasized the social role of tourism, which also gained ideological connotations: tourism and leisure should belong to the working people so they can gather new strength for further work (Rogoznica, 2014, p. 76; Kavrečič and Šuligoj, 2020). Even the most elite tourist destinations, such as Bled in the Slovenian Alps, were now dedicated to workers (Repe, 1996, p. 158). The members of the Yugoslav union organizations were granted a lower price for tourist accommodation and the annual allowance was granted by the Ministry of Finance (Rogoznica, 2014, p. 76). As stated by Duda: "Paid leave was a constitutional entitlement: employers had to accept it and employees had to take it." Although, the interference of the state in the tourism sector was not invented by the Yugoslav state, it goes back to the 19th century (Duda, 2010a, pp. 34–36; see also Judson 2002). As in previous cases (especially in totalitarian regimes in the interwar period), tourism was used by political ideology for the affirmation of its doctrine. In the case of socialist Yugoslavia, the creation of holiday homes for workers at affordable price was a clear sign of the efficiency and great accomplishments of the regime. Taking a holiday became an obligation. While enjoying the beauties of the country and "greatness" of the regime, the worker rested and regained strength for further work and the building of socialism.

Two years after the end of World War II, in 1947, the number of tourists in Yugoslavia had already reached the pre-war numbers. It is important to emphasize that out of almost 212,000 guests only 6,720 (3 %) were foreign tourists, which is ten times less than in 1938. Three years later the number of tourists was lower (107,786), but the ratio between domestic and foreign tourist slightly changed: 102,648 domestic (95 %) and 5,138 (5 %) foreign tourists. In 1953, the percentage of foreign tourists increased. 134,648 (89 %) domestic tourists and 16,425 (11 %) foreign were registered (Repe, 2006, pp. 64 and 67). These figures indicate significant oscillations in the recovery of tourism in the post-war period, reflecting well the difficult situation in the coastal area, state and in other war-affected emitive markets in Europe.

One of the reasons for the growth of post-war tourism was the construction of holiday homes – mainly trade union homes; this special type of social tourism was oriented towards social groups that had been overlooked in the past. Various companies and public sector organizations have built or opened holiday homes for their employees in order to grant them at least some gratification for their hard work throughout the year (Kavrečič and Šuligoj, 2020). The new system meant that workers, who could not afford to go on holiday in the past, engaged in tourism. One facilitator was the aforementioned paid leave and annual allowance, the other was the affordable price of accommodation. In fact, until the 1960s, trade unions largely organized tourism services for workers. Accommodation for an affordable price was also provided by the nationalization of private villas and hotels, adaptation of abandoned houses in the town centres or nearby surroundings with discounts for travel to tourist destinations for members of the trade unions. As previously mentioned, the attitude towards tourism changed in the 1960s, when the state turned the interest and focus towards commercial tourism and new hotel complex especially in Portorož, predominantly visited by foreign tourists (Repe, 1996; Duda, 2010a, pp. 36–38). Nonetheless, social and commercial tourism coexisted during the whole period of socialist Yugoslavia.

In the next section, we will highlight the narrow coastal area in Slovenia, which was one of the six constituent units of the Yugoslav Federation; explanation of the post-war development of the coastal accommodation sector and special emphasis given to the Municipality of Piran.

4 The accommodation sector within the new socialist ideology: The case of the Municipality of Piran

Between 1947 and 1954, the Municipality of Piran was part of Zone B of the Free Territory of Trieste (FTT) and was under the administration of the Yugoslav Army, which took over 34 accommodation establishments with a total capacity of about 1,400 beds. Nevertheless, the tourism industry began its real recovery when the armed forces abandoned two hotels – the Rotonda Hotel in Piran and the Central Hotel in Portorož – in 1947 (Brezovec, 2015, p. 121).[6] Within the new Yugoslav administration, the organization of tourism followed the "structural transformation of once mainly private activity into a predominant social sector." Although, due to the complex political situation in the FTT, some forms of adjustments during the transfer of ownership of tourist and catering facilities had to be taken into account (Rogoznica, 2005, pp. 395–396). Regardless of fairly modest demand, the new start, recovery and reorganization of tourist services quite successful. For example, in 1948 there were 170 beds available, in 1949 there were 440 and in 1950 there were about 1,200 beds in hotels and holiday homes (Rogoznica, 2005, p. 399). Moreover, at the end of 1951, Triglav Hotel was opened in Koper/Capodistria[7] as the first hotel constructed after the World War II. The investment cycle of the new socialist authority continued after the dissolution of the FTT, when six hotels with a total of 333 rooms were already operating on today's Slovenian coast: the Central, Helios and Palace hotels in Portorož, the Metropol hotel in Piran, the Triglav hotel in Koper and the Tourist hotel in Ankaran/Ancarano (Turistički savez Jugoslavije, 1954). Thus, large state-owned tourism companies first managed the nationalized tourism facilities and later built new ones. The greatest attention and support was given to Portorož, which was already established on the international environment as a tourist destination and one of the most important destinations in the Upper Adriatic. The growth of tourist demand and the need for foreign currency encouraged the Yugoslav authorities to permit and facilitate the crossing of state borders, accelerate investments in tourism and replacement of non-service activities from tourist centres (Brezovec, 2015; Šuligoj, 2015). Such centres were visited by both foreigners as well as domestic elites, who were, however, treated differently to foreigners.

In accordance with the new plan for the comprehensive development of the Slovene coast, a new regional plan was implemented. The main architect for that was Edo Mihevc, president of the new Urban Planning Council in the District of Koper. He designated a specific role to each of the coastal towns. Koper had become the administrative and industrial centre (due to the port opening in 1957), Izola/Isola the industrial and fishing centre. The main tourist role was assigned to Piran and Portorož (Čebron Lipovec, 2019, 206–207). In accordance with the general political agenda, in the first post-war years, the emphasis was

---

6   For tourism development in the zone B of FTT see also Rogoznica, 2005.
7   Koper was (is) a Municipality but also the capital of the district.

on social tourism (if we deliberately ignore the state's focus on the industry), later, the attention was given to international commercial tourism. Piran and Portorož made no exception. The major architect of the new holiday resorts on the Slovene coast within the *Regional plan for the Slovene Coast* from 1959 was also Edo Mihevc. The first holiday resorts were built in Strunjan/Strugnano, Ankaran, Lucija/S. Lucia and San Simon/San Simone, followed by hotel infrastructure in Lucija-Portorož and Izola. The peak was reached with the construction of the Metropol hotel and Bernardin resort in Portorož. In the coastal regional administrative centre, Koper, the only important hotel was the aforementioned Hotel Triglav in the town centre. Although, in the municipality area of the town, the Žusterna/Giusterna resort was built (Čebron Lipovec, 2018, pp. 210–211). These tourist accommodation facilities were mostly intended for foreigners and domestic elites. However, in accordance with the Yugoslav ideological social project, workers' tourism, which today would be understood as trade union or even social tourism, was still prioritized. In 1958, 68 holiday homes from various places in Slovenia were opened in the Municipality of Piran. Abandoned buildings were arranged for the purpose of accommodation for domestic tourists. The fact that holiday homes were intended for domestic tourists and hotels for foreign and elite tourists is made clear by the following data: in 1963 33,479 visits were recorded in holiday homes in the area of Piran, of which 31,891 were domestic guests. In hotels in the same area, the situation was exactly the opposite. A total of 1,220 guests were recorded, most of which, 1,182 were foreign. The share of domestic guests in hotels was almost negligible (SI PAK PI 36; Kavrečič et al., 2019). Accordingly, in the 1950s the post-war coastal tourism depends on domestic tourism. Later in the 1960s, commercial tourism became more important. In the Croatian part of the Adriatic, for example, commercial tourism prevailed due mainly to the introduction of holiday allowance and the abolition of benefits associated with public transport on holiday (Duda, 2010, p. 294; Duda, 2015, p. 53).

Holiday homes of various state-owned companies and organizations were a special type of accommodation opened all along the Slovenian coast.[8] The social tourism for workers developed especially in the 1950s, when 36 holiday facilities in FTT were given to companies and institutions of the People's Republic of Slovenia in order to provide accommodation for workers during their holiday. After 1954 and the dissolution of the FTT, the Yugoslav authorities encouraged the further development of social tourism for the working people. This can be

---

8   And also elsewhere along the Yugoslav coast in the Eastern Adriatic.

asserted also by the growing number of holiday homes. In 1954, there were 9 homes in the Koper district, in 1955 the number increased to 12 and in 1956 to 56, which was supposed to prove that "the Slovenian coast began to serve not only international tourism, but also the domestic working man as a place of rest and relaxation." In 1954 there were 9 holiday homes in the district of Koper, in 1955 12 and in 1956 already 56 (Rogoznica, 2014, p. 77). The greater investment in accommodation infrastructure for workers took place in the Municipality of Piran. In fact, in the newspaper Slovenski Jadran, 68 holiday homes in Piran were recorded in 1958 (Slovenski Jadran, 1958). Four years later, in 1962, there were already 110 holiday homes in the whole Municipality of Piran (82 holiday homes in Piran-Portorož and 28 in the area of Piran-Fiesa) with a total of 3,880 beds (2,635 beds in Piran-Portorož area and 1,245 beds in Piran-Fiesa area) (SI PAK PI 36). The companies that owned the homes were from all over Slovenia, all economic sectors, as well as public sector, were also represented (SI PAK PI 36; Kavrečič et al., 2019). Their number increased rapidly, and just before the disintegration of socialist Yugoslavia, there were over 7,100 beds in holiday homes and youth holiday camps on the Slovenian coast (Brezovec, 2015, p. 122). They had fairly basic accommodation units (rooms) available to workers at a non-commercial price (cost price). The equipment was very basic and offered no comfort; usually only beds for two or more people (family rooms), bedside tables (or shelves next to the bed), and closets. Rooms often did not have private bathrooms; guests used larger shared bathrooms with basic sanitary facilities located on each floor. All guests were obliged to adhere to peace regulations (from 10 pm till 6 am), order and cleanliness of the home. Bedding was not provided in the homes. All guests were in charge of the tidiness of their room with the common areas, such as bathrooms, hallways and stairways being cleaned by the cleaning lady (SI PAK PI 36.726). However, hygiene was often a weak point in these facilities and thus the subject of comments and complaints. Many holiday homes were provided by a kitchen, which offered half or full board and beverages also at a cost-price. All homes submitted monthly and seasonal reports on overnight stays, the number of tourists, and the income obtained from overnight stays and catering services. Each home had a house rules and operating instructions. The homes were inspected where possible deficiencies (installation of fire extinguishers, renovation of the room, replacement of furniture, cleanliness and tidiness) were pointed out (SI PAK PI 36.726). The rules (*pravilnik*) also determined the ownership of the home and all its furnishing, which belonged to all workers of that specific company.

The main investors ("owners") in worker's tourism were the Slovenian continental state-owned companies and the ministries for public sector employees.

Slovenian investors invested not only on the Slovenian coast, but also elsewhere in the eastern part of the Upper Adriatic, including the rest of the Istrian peninsula, Kvarner (also on the islands) and further south. Holiday homes were not classified and labelled according to quality levels. Facilities and services offered were not uniform and depended, of course, on the companies to which they belonged and on the creativity of the caretakers and the union. Investors/owners themselves were then responsible for the management and maintenance of these holiday facilities. Workers arranged their holiday dates with the trade union, who in practice acted as a booking service. The services were then charged directly by the employer (property owner); the employer had the right to made a deduction from the worker's pay. The use was not treated as a fringe benefit and therefore not taxed (no tax on bonuses). The investors in agreement with the union employed a caretaker to organize work during the home's operation, e.g., maintenance work, keeping records of guests and property, serving food and drink (if offered) and the like.

The guest books of the holiday homes owned by different organizations and companies provide us with information about the visitors: name, last name, date of arrival and department, place of living, occupation, marital status (Kavrečič et al., 2019; Kavrečič and Šuligoj, 2020). The results of a preliminary analysis of 19 guest books for the period of 1956 to 1965 show that the vast majority in the holiday homes were domestic guests (98 %). The tourist season in most cases lasted from May to September. In one season, between 300 and 600 guests spent their holidays in each home, staying there for an average of 9 days (SI PAK PI 36). Hence, if consider 1962, when 110 holiday homes were situated in the Municipality of Piran, we can note they were owned by different companies from all over the Socialist republic of Slovenia. Most of the homes' owners came from the secondary sector (46 or 42 %), followed by the quaternary sector (39 or 35 %), tertiary (15 or 14 %) and finally the primary sector (10 or 9 %). Most of the home owners were companies from the capital, Ljubljana. This is understandable, since the capital was the economic and political centre of the republic. The companies from the city cover 48 % of the accommodation for workers in the Municipality of Piran. The other two most industrialized regions (on the northeast and north) followed with 18 % and 15 %. Other regions followed with less than 10 % of holiday homes.

In addition to the practices mentioned above, it is necessary to refer to other types of accommodation that can be placed in the context of workers' tourism, e.g., apartments, bungalows, campsites (individual pitches in campsites). They were also owned or merely leased by state-owned enterprises. The management was similar to that of holiday homes.[9] As a kind of systemic exception, which was not in line with the socialist mentality of Yugoslavs about general "brotherhood and unity," and equality, there were second or private holiday homes or holiday cottages.[10] The part of the population (0.7 % of the total non-agricultural population) that was more educated, had a better income and social status, and had a more pronounced need for living and recreation in nature (residents of urban areas) invested in their own second homes, especially after 1961. In 1967, there were 1,895 cases in the Slovenian part and 1,131 in the Croatian part of western Istria (Jeršič, 1968).[11] Private second homes provided socialist elites with quality leisure time away from the working class masses. In addition, these elites could also stay in hotels at special rates, which were too high for the domestic working class.

Finally, let us summarize that "socialist-style tourist accommodation" was actually a very inhomogeneous coinage that included very different accommodation establishments (i.e. private holiday homes and those for workers, hotels, apartments, bungalows, campsites (individual pitches in campsites)), which were also used differently by individual groups within socialist society.[12] The Yugoslav socialist myth of equality and solidarity was certainly violated in the sphere of holiday practices. The development of the "socialist man" (as derive also from the slogan mentioned in the introduction) was therefore not as systematically uniform as might be expected. Variations are apparent even in the examination of so narrow an area as leisure and holiday. Although these practices can also be discussed from a broader field of political science[13] or tourism – political connections, we have only mentioned them as this is beyond the purpose of this chapter.

---

9   In the campsites, this was coordinated with the campsite management.
10 These are just different denominations for the same private property, *vikend* in Slovenian.
11 It is interesting to note that the author attributes the increase in these properties also to the abandonment of agricultural activity and the move of people to the cities (rural depopulation), but does not report on the abandoned properties of the Italian emigrants (*esuli*). More on these private properties can be found also in Taylor (2010).
12 In this context, although separately, foreign tourists should also be considered. We have purposely not addressed this segment in this section.
13 More can be found in Mihaljević (2019).

5 Conclusion

This chapter sheds light on the development of a specific form of tourism that began to gain an importance on the Eastern side of the Upper Adriatic after World War II (and the socialist revolution), or more precisely after the resolution of the Italian-Yugoslav border issue. In this context, this research is an extension of Kavrečič's work (2017) explaining the development of coastal tourism during the Austro-Hungarian Empire. Moreover, it also an extension of Šuligoj and Sinkovič's (2018) paper focused on the turbulent period between the two World Wars. Historians of this period usually investigate other, non-tourism issues, typically related to fascism. The same applies to the post-World War II period, when the border issue, mass migrations, the new socio-political system and similar topics are at the forefront. Accordingly, the recovery of the tourism industry in the post-World War II period, within the framework of the new socialist mentality on the territory of present-day Slovenia, was thus a somewhat overlooked research topic. Thanks to some historians from Pula/Pola (Croatia), Istria and other regions in Croatia are certainly better researched in this context. Nevertheless, some authors such as Blažević (1984) and Šuligoj (2015) made an overview of the development of tourism in the region. Together with Grandits & Taylor's (2010) monograph, they certainly explain the development of tourism and the broader socio-economic context during the period of socialist Yugoslavia. However, some micro-locations (destinations) and their specificities require additional investigations and explanations; the same applies to research on service providers. Thus, the contribution of this chapter lies precisely in the fact that it examines the accommodation industry in a specific geographical micro-area, in a specific period in the second half of the 20th century.

Authors highlighted the specific Yugoslav model of worker's tourism and related "socialist-style tourist accommodation" within domestic tourism, focusing on the case of the Municipality of Piran, which has become the centre of domestic seaside tourism in Slovenia. The chapter highlights only a period or phase of development in the significant metamorphosis of Slovenian coastal tourism in the studied area. In the post-war period, a conceptual shift took place in the area from elite tourism, conceived during the Habsburg era, to social tourism, aimed at the working class. However, the Yugoslav socialist model based on the working class was obviously not entirely consistent as elites nevertheless holidayed (rested and relaxed) differently than the working class. In practice, this means the coexistence of two models, one based on the left social ideas and the other was clearly more similar to those of Western capitalists; models did not include the same type of accommodation. This would otherwise be nothing special in the context of Western plural/democratic societies and capitalism where low-income workers were, however, often overlooked. This Yugoslav social responsibility was reflected only in relation to domestic workers and their holidays, while no solidarity was evident towards foreign workers. The international political slogan "Workers of all lands, unite!" was obviously not included in the international tourist flows, although everywhere in Europe this was associated with the acquired right to a holiday (rest and relax) for the working class. In the Yugoslav case, this meant that foreign workers as tourists were in a worse position than domestic ones and were understood as a source of foreign currency in the context of commercial tourism.

6 Sources

Pokrajinski arhiv Koper/Regional Archive Koper, unit Piran.

SI PAK PI 36. Skupščina občine Piran s predhodniki. Počitniški domovi 1962-1965 (odločbe). Knjige gostov počitniških domov OBLO Piran.

SI PAK PI 36.726, Pravilnik o organizaciji in poslovanju počitniškega doma v Piranu, 1959.

References

Ashbrook, J. (2006). "Istria Is Ours, and We Can Prove It": An Examination of Istrian Historiography in the Nineteenth and Twentieth Centuries. *The Carl Beck Papers in Russian and East European Studies* (1707), 1–39.

Blažević, I. (1984). *Turizam Istre*. Zagreb: Savez geografskih društava Hrvatske.

Brezovec, T. (2015). "Razvoj hotelske ponudbe na slovenski obali." In: M. Šuligoj (Ed.), *Retrospektiva turizma Istre* (pp. 111–142). Koper/Capodistria: Založba Univerze na Primorskem.

Bufon, M. (2008). "Razvoj in struktura območij družbenega in kulturnega stika v zgornjem Jadranu." In: M. Bufon (Ed.), *Na obrobju ali v osredju? Slovenska obmejna območja pred izzivi evropskega povezovanja* (pp. 61–139). Koper: Univerza na Primorskem, Znanstveno-raziskovalno središče, Založba Annales; Zgodovinsko društvo za južno Primorsko.

Bufon, M. (2009). "Zgornji Jadran: prostor konflikta ali koeksistence?." *Annales. Series historia et sociologia* (19, 2), 457–468.

Čebron Lipovec, N. (2018). *Izgradnja slovenskih obalnih mest v času po drugi svetovni vojni: primer mesta Koper: doktorska disertacija (PhD thesis)*. Koper: Univerza na Primorskem, Fakulteta za humanistične študije.

Čebron Lipovec, N. (2019). "Post-war urbanism along the contested border: some observations on Koper/Capodistria and Trieste/Trst." *Dve domovini: razprave o izseljenstvu* (49), 199–220.Duda, I. (2010a). "Adriatic for All: Summer Holidays in Croatia." In B. Luthar & M. Pušnik, (Eds.), *Remembering Utopia: The Culture of Everyday Life in Socialist Yugoslavia*. Washington (pp. 289–311). DC: New Academia Publishing.

Duda, I. (2010b). "Workers into Tourists: Entitlements, Desires, and the Realities of Social Tourism under Yugoslav Socialism." In H. Grandits & K. Taylor (Eds.), *Yugoslavia's Sunny Side. A History of Tourism in Socialism (1950s–1980s)* (pp. 33–68). Budapest-New York: Central European University Press.

Duda, I. (2015). "Socijalni turizam i socijalizam: slučaj Fažana." In: M. Cerić & M. Mrak (Eds.), *Zbornik javnih predavanja 3* (pp. 49–58). Pazin: Državni arhiv u Pazinu.

Flere, S., (2012). "Da li je Titova država bila totalitarna?." *Političke perspektive*, (2, 2), 7–21.

Hrobat Virloget, K. (2015). "The burden of the past: silenced and divided memories of the post-war Istrian society." In K. Hrobat Virloget, C. Gousseff & G. Corni (Eds.), *At home but Foreigners: Population Transfers in 20th Century Istria* (pp. 159–187). Koper: University of Primorska, Science and Research Centre, Annales University Press.

Hrobat Virloget, K. (2021). *V tišini spomina. "Eksodus" in Istra*. Koper, Trst: Založba Univerze na Primorskem, Založništvo tržaškega tiska (in print).

Jeršič, M. (1968). "Sekundarna počitniška bivališča v Sloveniji in Zahodni Istri." *Geografski vestnik* (40), 53–67.

Judson, P. (2002). "Every German Visitor Has a Völkisch Obligation He Must Fulfill. Nationalist Tourism in the Austrian Empire, 1880–1918." In R. Koshar (Ed.), *Histories of Leisure* (147–168). Oxford and New York: Berg Publishers.

Kavrečič, P. (2017). *Turizem v Avstrijskem primorju. Zdravilišča, kopališča in kraške jame (1819–1914)*. Koper: Založba Univerze na Primorskem.

Kavrečič, P. (2020). "Tourism and fascism: tourism development on the eastern Italian border." *Prispevki za novejšo zgodovino* (60, 2), 99–119.

Kavrečič, P. & Radošević, M. (2017). "Na morje! Izzivi turističnega razvoja v Istri v času Avstro-Ogrske in italijanske uprave s posebnim ozirom na leto 1925." *Zgodovina za vse, vse za zgodovino* (24/2), 21–40.

Kavrečič, P. & Šuligoj, M. (2020). "Post-World War II seaside workers' holiday homes in the Eastern Adriatic. The case of Piran/Pirano." *Convegno internazionale Verso la massificazione. Il turismo nell'area euro-mediterranea: politiche, società, istituzioni ed economia*, svolto a Napoli nei giorni 1 e 2 ottobre 2020.

Kavrečič, P. et al. (2019). "Počitnikovanje delavskega razreda, prezentacija praks iz obdobja socialistične Jugoslavije na primeru občine Piran": elaborat projekta (razstave). Koper: Univerza na Primorskem.

Kosmač, M. (2017). *"Etnično homogena Evropa": preselitve prebivalstva v Istri in Sudetih 1945-1948.* Koper: Znanstveno-raziskovalno središče, Založba Annales.

Marcks, J., Knieling, J. & Vladova, G. (2016). "The North East Adriatic Region: Territorial Cooperation and the Role of Planning Systems and Cultures." *European Spatial Research and Policy, The Journal of University of Lodz* (23, 2), 5-24.,

Medica, K. (2011). "Prekinitve in diskontinuitete razvoja, stalnica istrske obmejnosti." In: D. Darovec & P. Strčić (Eds.), *Slovensko-hrvaško sosedstvo: Hrvatsko-slovensko susjedstvo* (pp. 249-262). Koper: Univerza na Primorskem, Znanstveno-raziskovalno središče, Univerzitetna založba Annales: Zgodovinsko društvo za južno Primorsko.

Mihaljević, J. (2019). "Social Inequalities from Workers' Perspective in 1960s Socialist Yugoslavia." *Revue d'études comparatives Est-Ouest* (1,1), 25-51.

Nejamšić, I. (2014). "Iseljavanje iz Hrvatske od 1900. do 2001.: demografske posljedice stoljetnog procesa." *Migracijske i etničke teme* (30, 3), 405-435.

Oblak Moscarda, O. (2016). *Il "potere popolare" in Istria 1945-1953.* Rovigno: Centro di ricerche storiche; Fiume: Unione Italiana; Trieste: Universita popolare di Trieste.

Pipan, P. (2007). "Cross-border cooperation between Slovenia and Croatia in Istria after 1991." *Acta Geographica Slovenica* (47, 2), 223-243.

Pirjevec, J. (2007). *"Trst je naš!" Boj Slovencev za morje (1848-1954).* Ljubljana: Nova revija.

Purini, P. (2012). "Esodi dalla Venezia Giulia: cause politiche e motivazioni sociologiche." *Acta Histriae* (20, 3), 417-432.

Repe, B. (1996). "Turizma ni mogoče zavreti, čeprav bi ga prepovedali z zakonom. In F. Rozman & Ž. Lazarevič (Eds.), *Razvoj turizma v Sloveniji: zbornik referatov* (pp. 157-164). Ljubljana: Zveza zgodovinskih društev Slovenije

Repe, B. (2006). "Turistična zveza in razvoj turizma v Sloveniji po drugi svetovni vojni." In S. Šajn (Ed.), *Turizem smo ljudje: zbornik ob 100-letnici ustanovitve Deželne zveze za pospeševanje prometa tujcev na Kranjskem, Turistične zveze Slovenije in organiziranega turizma v Sloveniji: 1905-2005* (pp. 61-99). Ljubljana: Turistična zveza Slovenij.

Reverdito, R. (2009). "Land and sea boundaries between Slovenia and Croatia from federal Yugoslav,ia to the Schengen fortress." In: M. Sobczyński (Ed.), *Historical regions divided by the borders. General problems and regional issue* (pp. 59-68). Łódź-Opole: University of Łódź, Department of Political Geography and Regional Studies & Governmental Research Institute, Silesian Institute in Opole & Silesian Institute Society.

Rogoznica, D. (2005). "Obnova in razvoj turizma na območju cone B Svobodnega tržaškega ozemlja (s posebnim ozirom na okraju Koper)." *Acta histriae* (13), 395–422.

Rogoznica, D. (2014). "Aplikacija in uveljavitev modela socialističnega turizma na slovenski obali (1947–1990)." *Arhivi* (37, 2), 73–84.

Šarić, T. (2015). "Bijeg iz socijalističke Jugoslavije – ilegalna emigracija iz Hrvatske od 1945. do početka šezdesetih godina 20. stoljeća." *Migracijske i etničke teme* (31, 2), 195–220.

Šuligoj, M. (Ed.) (2015). *Retrospektiva turizma Istre*. Koper: Založba Univerze na Primorskem.

Šuligoj, M. & Sinkovič, L. (2018). "Ladies and gentlemen, all aboard!" Travelling in the northern Adriadic until the World War II. *Historický časopis* (66, 4), 629–648.

Taylor, K. (2010). "My Own Vikendica: Holiday Cottages as Idyll and Investment." In H. Grandits & K. Taylor (Eds.), *Yugoslavia's Sunny Side. A History of Tourism in Socialism (1950s–1980s)* (pp. 171–210). Budapest-New York: Central European University Press.

Taylor, K. & Grandits, H. (2010). "Tourism and the Making of Socialist Yugoslavia. An Introduction." In H. Grandits & K. Taylor (Eds.), *Yugoslavia's Sunny Side. A History of Tourism in Socialism (1950s–1980s)* (pp. 1–30). Budapest-New York: Central European University Press.

Troha, N. (2018). "Ustvarjanje meje z Italijo in vloga popisov prebivalstva." In M. Zajc (Ed.), *Ustvarjanje slovensko-hrvaške meje, Vpogledi 19* (pp. 165–188). Ljubljana: Inštitut za novejšo zgodovino.

Troha, N. (2019). "Nekaj utrinkov iz političnega življenja na Svobodnem tržaškem ozemlju (1947–1954)." *Kronika: časopis za slovensko krajevno zgodovino* (67, 3), 677–692.

Turistički savez Jugoslavije (1954). *Hotelski vodič. Jugoslavia*. Beograd: Turistički savez Jugoslavije.

Violante, A. (2009). "The past does not seem to pass at the Venezia Giulia border." In: M. Sobczyński (Ed.), *Historical regions divided by the borders. General problems and regional issue* (pp. 97–112). Łódź–Opole: University of Łódź, Department of political geography and regional studies & Governmental research institute, Silesian institute in Opole & Silesian institute society.

Žerjavić, V. (1993). "Doseljavanja i iseljavanja s područja Istre, Rijeke i Zadra u razdoblju 1910–1971." In: *Društvena istraživanja: časopis za opća društvena pitanja* (2, 4–5, 6–7), 631–656.

Tomi Brezovec and Aleksandra Brezovec

Yugoslavia awaits you: post WW2 tourism promotion of the Yugoslav coast

Abstract The purpose of this chapter is to discusses the development of tourism promotion on the eastern coast of the Adriatic during first decades after the WW2. After the war, the Istrian peninsula and the whole Dalmatian coast became part of Yugoslavia, a country that has established a new socialist political realm. In the two decades following the war, the government efforts were first concentrated on rebuilding the infrastructure and in development of the industry and tourism was not a priority. However, with the evolvement of tourism demand in Europe, the government started to consider international tourism as a significant source of foreign currency, needed for development of the economy. The shift from domestic workers tourism to foreign tourist generating markets is clearly visible through tourism promotion. The content analysis of official tourism promotion material has revealed significant changes in textual as well as in pictorial/photographic representations of coastal tourism destinations. The shift in governmental support of international tourism laid the foundations for mass tourism that has developed in the following decades.

1 Introduction

Contribution of tourism to the balance of payments, generating jobs, or regional development are some of most evident and widely acknowledged economic benefits of tourism. Being not only economic but also a social phenomenon tourism can be used in pursuit of non-economic goals, such as support of specific political ideas, or strengthening of a nation's unity (MacCannell, 1992). Tourism promotion can therefore be used in pursuing both, economic and non-economic benefits of tourism. It can be used to communicate general or specific message to selected market segments and it can pursue general or market specific interests. Tourist brochures, folders and tourist maps are considered as most effective promotion publications in tourism (Čulić, 1965). The use of brochures grew with the development of tourist demand and with increased competition on tourism market. They are cheaper to produce than tourist guides and monographic books and can be easily distributed in large quantities. Their content can also be easily adapted to specific markets or to changes in tourism product. As such they are a good source of information and can provide an insight into past tourism promotion policies.

In this chapter, the case of post-war tourism recovery and development of tourism in Yugoslavia illustrates how tourism promotion was used to support government's political agenda as well as to foster tourism's economic benefits. The study of official travel promotion publications shows that they are a good reflection of Yugoslav tourism policy and its priorities. The analysis of textual content and design of brochures shows also gradual shift in government's focus from social and political to economic benefits of tourism that occurred during post-war recovery period.

2 Tourism development on the eastern coast of the Adriatic

Tourism on the eastern coast of the Adriatic started during the 19th century, when the area was part of the Habsburg Empire that collapsed in 1918, at the end of the first world war. Contrary to inland spa and watering destinations that practiced tourism since the 18th or early 19th century, coastal tourism gained its momentum only during the last quarter of the 19th century. Relatively late development was due to perceived unattractiveness of the coastal environment. Port activities, shipbuilding, fishing, or salt harvesting were not an attractive environment for those who could afford to travel for pleasure. Also, coastal climate and sea water were considered unhealthy. By the end of the century, the advances in medical science confirming the benefits of sea-bathing, as well as the development of seaside tourism in other parts of Europe, contributed to the shift in perception of the coastal environment. Consequently, during the last two decades of the 19th century, the coast has become attractive to investors who built holiday villas, hotels, restaurants, and other tourism infrastructure. In about two decades before the outbreak of the first world war, the north-eastern Adriatic coast that was closer to the capital and relatively easily accessible, evolved from being unknown and undesired place into new Austrian Riviera. Some coastal destinations became very fashionable among members of the upper class. Abbazia (today Opatija in Croatia) was one of the most visited destinations in the Monarchy, second only to Karlsbad (today Karlovy Vary in Czech Republic).

The outbreak of the WWI has interrupted fast growth of tourism. Many planned or already started investments were abandoned (Kranjčević, 2019), while the existing tourism infrastructure was put in military use. Especially large hotels and sanatoria were usable as hospitals and recovery centres. After the war, tourism reassumed in a completely different political environment. After the collapse of the Habsburg Empire, the coast was divided between Italy and the newly formed Kingdom of Serbs, Croats, and Slovenians. Italy got the northern part of the coast (the former Austrian Riviera), consisting of Istrian peninsula and parts of the Gulf of Quarnero with some islands, and of the town of Zadar on Dalmatian coast. The rest of the coast was included in the newly formed Kingdom of Serbs, Croats, and Slovenians, later renamed the Kingdom of Yugoslavia. Changes of political borders had a strong impact on tourism. Changes were evident in the social structure of visitors, in the number of tourists, and in their way of spending holidays at the sea. While during the pre-war period, only the upper class could afford seaside holidays, the post-war period introduced the beach to the middle-classes. Sea bathing has become more popular and demand for amusement activities has also grown compared to pre-war tourism demand. Spending the day at the beach, relaxing, sunbathing and swimming has become a new norm. In the Kingdom of Yugoslavia, tourism has seen a significant growth (Benić Penava & Matušić, 2012). Several new coastal destinations have emerged, increasing the competitivity on the market. Especially destinations on the middle and southern Dalmatian coast, and on the coast of Montenegro, that were previously distant from main tourist generating markets of Austria and Bohemia, have benefited from increased demand from the new domestic market that was much closer. Besides the coast, inland spas in Slovenia, Croatia, Bosnia, and Serbia also gained in popularity. Another slow-down in tourism development took place at the beginning of 1930s, as a consequence of the 1929 world economic crisis. Tourism investments as well as tourism demand have diminished, and many businesses had to close their activities for a couple of years or went out of business. The recovery period after the crisis was short-lived due to the outbreak of WW2 and a direct attack of Axis Forces on the Kingdom of Yugoslavia in April 1941.

3 Post-WWII tourism recovery

Immediately after the Second World War, the eastern coast of the Adriatic was again subject to political changes. The entire coast became part of Yugoslavia, except for a small area around the town of Trieste and part of northern Istria, where the border with Italy has been agreed only in 1954. The post-war Yugoslavia had also introduced a socialist political system, where workers were considered pillars of the society. These changes had a strong impact on tourism. Before the establishment of the socialist state, tourism was considered and activity of the middle and upper classes of the society. With changed paradigm, workers, now considered a key element of the society, gained the right to rest to "regain strengths needed for the construction of new socialist society." The government has put all its efforts in rebuilding of the country and in the industrialization of its economy (Allcock, 1986). Tourism and leisure were used more as means for building new society, and to strengthen the "brotherhood and unity" among the peoples of Yugoslavia, rather than being regarded as profitable economic activity. Social, ideological, and political importance of tourism were therefore considered as important, if not more, than were its economic benefits (MacCannell, 1992; Yeomans, 2010). Trade unions and children colonies had the priority in use of existing tourism infrastructure. Even in most elite destinations, such as Bled, one of pre-war Yugoslav royal family holiday retreats, lodging capacities were supposed to be used exclusively for the holiday of workers (Repe, 1996). Foreign tourists were of no priority at all. A small number of foreign tourists that did visit Yugoslavia during the immediate post-war years were visitors from East European countries, mostly from Czechoslovakia and Hungary (Stanković, 1990). In 1948, there were 61,500 foreign visitors with a 4 % share in all tourism arrivals, compared to 1,617,000 domestic tourists (96 %). This small number of foreign visitors has been halved in the following year as a result of a 1948-dispute between Yugoslavia and Moscow. Yugoslavia was expelled from Moscow led organization of European Communist Parties. Consequently, the Soviet Union and other East European countries broke off relations with Yugoslavia. Tourism was no exception. The fall in foreign visitor numbers was more than outset by a 35 % growth of domestic tourists in 1949, and by another 16 % in 1950, when they hit 2.3 million mark.

After the break with Eastern European block, and with neglected relationship with the West, Yugoslavia needed a way to reposition itself on political and economic map (Tchoukarine, 2015). Already in 1948, immediately after the break with the Soviet Union, Yugoslavia started a "come and see the truth" campaign with the aim to attract foreign leftist intellectuals from the West to visit the country and became a testimony of country's progress (Tmušić, 2013; Tchoukarine, 2015). On top of that, tourism started to be considered as a possible source of convertible foreign currency that was needed to finance imports from West European countries and from the USA (Allcock, 1986).

4 Early post-war tourism promotion

The break with the East European block has stimulated tourism orientation towards West European countries (Tchoukarine, 2010; Yeomans, 2010; Tmušić, 2013). One of the first publications targeting foreign tourist was published in 1949. A 9x15 cm colour four-page folder printed in German entitled "*Besucht Jugoslawien*" and French "*Visitez la Yougoslavie*" version, merely listed four destinations on "The Yugoslav Riviera" – Opatija, Rab, Split and Dubrovnik, providing limited information on prices, visa system and foreign exchange policy. In the same year, another brochure was produced bearing the title FNR Jugoslavija (Federal People's Republic of Yugoslavia). A colour front cover with a drawing of a women in traditional costume and a sketch of Dubrovnik's landscape, followed by 16 folded full page sepia photographs of 16 tourist destinations, from Rovinj in the north, following the coast, to Hercegnovi in Montenegro, with no other text, besides place names. So designed, the brochure could be used for foreign as well as for domestic promotion of the Yugoslav coast. Apart of historic coastal towns not much else was available to offer to foreign tourists in terms of tourist attractions. A half-century of tourism development on the Eastern Adriatic coast was hit by two wars and an economic crisis. After the WWII, the existing infrastructure was not up to expected foreign standards. Tourism, still being a relatively low investment priority, was limited to upgrading and modernization of existing facilities which were, in line with the political agenda, primarily used to provide accommodation for socialist workers.

The government was more interested in domestic tourism and its use for ideological purposes. Tourism was considered a tool to be used in creation of a "new Yugoslav citizen" (Yeomans, 2010, p. 72). Local population was not used to take holidays and to travel for leisure purposes, and in an effort to turn "workers into tourists," the government introduced a number of entitlements and financial incentives such as paid leave and travel subsidies, or provision of a network of holiday centres for workers (Duda, 2010; Sitar, 2020). The efforts in fighting populations' tourism "illiteracy" and educating citizens about positive effect of regular daily, weekly, and yearly rest on labour productivity, can be recognized in tourist brochures targeting domestic market. A 1949 16-page brochure issued by Putnik, a state travel agency, bearing the title "Workers, take holidays!" starts with text that reads:

> The New Yugoslavia has ensured its workers, fighters for the construction of socialism, the right to paid leave. The fulfilment of numerous tasks in the process of our construction of socialism depends on adequate use of yearly leave. Well used time-off refreshes and regenerates strengths needed for work efforts. Therefore, the use of yearly leave is not only a right but is also mandatory for our working people. (Putnik, 1949; translation by the author).

The government introduced a two-week annual paid leave already in 1946 (Službeni list, 1946). This was well above six days suggested by the International Labour Organization in 1936 Holidays With Pay Convention (Convention 52). In subsequent years this period was extended several times, to reach a minimum of 18 and maximum of 30 day paid leave by 1973 (Službeni list, 1973). To stimulate workers to travel to tourism centres and other parts of the country, the government introduced additional measures such as 50 % reduction on trains and coaches, or reductions in accommodation prices (Duda, 2010). The system, however, has not produced expected results. Workers were often reluctant to spend holidays in pre-selected destination by a trade union, or to spend holidays without their family members. In rural areas, workers preferred to stay at home and work on their fields rather than spend two weeks in a, potentially, undesired destination. In line with the political slogan of "brotherhood and unity," summer holidays were organized in a way that workers from one of the federal republics would spend time in another republic, or that people from the rural area would visit urban centres and vice versa (Yeomans, 2010; Duda, 2016). However, most workers holidays that took place were spent at the seaside. One of the reasons was that there were at least some acceptable lodgings available in existing pre-war hotels and holiday villas. It was also easier to persuade workers to take summer holidays at the coast rather than on a countryside or in urban centres. For many people this was also a unique opportunity to see the sea for the first time. Lakes, mountain resorts and spas were other attractive options, but they could not compete with the attractivity of seaside destinations.

Information booklets published by Trade Union Association of Yugoslavia in 1950 (*Domovi odmora*) and 1951 (*Radnička odmarališta Centralnog odbora SSJ*) continued with the justification of a yearly holiday break, emphasizing the benefits for the individual, as well as for the socialist society. They also point out the benefits of the socialist system for workers, such as is the use of former upper-class holiday facilities for holidaying of workers and farmers. The brochures listed some of the most attractive holiday facilities and provided some statistics over the number of available holiday homes opened in past years. There were 18 holiday homes available in 1946 while by 1950 their number already exceeded 150. Similar brochures were published also in following years, growing in volume as a result of the fast growth in available holiday facilities for workers. Throughout 1960s, the Tourist Association of Yugoslavia published yearly a two-volume directory *Kuda na odmor* (Where to go for holidays), one volume listed seaside destinations (Primorska mesta) and one for mountain, lake, and spa resorts (Planine, jezera i banje).

5 Orientation towards western tourism markets

After the war, the international tourism in Europe was limited due to infrastructure conditions and the post-war standard of living. Many countries have also restricted their citizens from traveling abroad and encouraged them to spend their holidays, and their money, in their domestic resorts. By the end of 1940s, the economic situation has improved, and international tourism started its postwar growth. During early 1950s the Yugoslav government's attitude towards foreigners visiting the county has changed. The "come and see the truth" campaign had ideological background and was used to promote travel to Yugoslavia to enable visitors to experience the new socialist reality. Foreign visitors were considered witnesses of socialist reality, and that was politically more important than was their potential economic contribution to Yugoslav economy (Tchoukarine, 2010). During the 1950s, after Yugoslavia alleviated the threat of political isolation and managed to consolidated country's position on political map, the government started to put more weight on economic advantages of tourism, the position that has eventually prevailed during the 1960s on the expense of its tourism's cultural and political benefits (Tmušić, 2013).

The economic and political reforms of 1952 ended the period of "administrative management of economy" during which state policies directed all activities, including tourism (see Radišić, 1981; Kobašić, 1987). Decentralization of the economy included the dissolution of PUTNIK, a state travel agency, which had the exclusive right to deal with foreign tourism companies, and transformation of its regional offices in each of the republics into independent commercial travel agencies. They started to compete on the market for domestic and foreign customers and were given the right to establish commercial relations with foreign partners (Tchoukarine, 2015). Devaluation of dinar made Yugoslavia cheap for foreign visitors and the response from foreign markets was quick. The number of foreign tourists grew from 70,000 in 1951 to almost 130,000 in 1952, having a 4.9 % share of 2.6 million of the total tourist arrivals recorded in Yugoslavia that year. The year of 1952 is also considered as the beginning of foreign tourism development in Yugoslavia (Pirjevec, 1988). Another step towards internationalization of tourism was the establishment of Yugoslav national tourism office, the *Tourist Association of Yugoslavia* (Turistički savez Jugoslavije) in April 1953. Among its tasks were the promotion of Yugoslavian tourism on foreign markets, and the coordination of representative offices abroad (Čulić, 1965). The Association immediately established a network of representative offices in Western European cities – Vienna, Frankfurt, Paris, Stockholm, London and in New York. Later, offices opened also in The Hague, Zurich, Rome, and Athens. It also immediately started the production of tourism promotion materials. It should be noted here that promotion of tourism destinations and promotion of culture and nature of Yugoslavia was not without political connotation. The decision to open to Western European tourists was accompanied by a requirement to inform foreign visitors not only about the beauty of the country but also about the socialist system and its achievements, as well as about heroism of Yugoslav people in the National Liberation Struggle during the WWII (Yeomans, 2010).

Tourist Association of Yugoslavia started immediately with the production of promotion materials. Already in 1953 it produced a publication, a hybrid between a brochure and a tourist map, in several language versions of which two were in English – separately for the UK and the USA market. All brochures had the same graphic layout and identical green coloured photographs. The colour front page had four generic drawings of a camping site, of a mosque, of a church, and of a spa spring. The title invited the reader to visit Yugoslavia – *Yugoslavia awaits you* (UK edition), *Yugoslavia is waiting for you* (USA edition), *Jugoslawien ruft* (German), *Yougoslavie vous attend* (French). The textual content in brochures was different in each language version. The English text for the UK edition differed from the USA edition in that only described the land and its people, the Adriatic, historical monuments, and folklore, and provided some information on visa and other vital information for visitors. The USA edition text was a combination of a political pamphlet and a tourist guide. The text literally begins with the explanation that "Yugo" means "south" and therefore Yugoslavia means "the land of south Slavs." It describes multi-ethnic and multi-religious composition of the country, and presents basic information on the constitution, foreign trade, and the relationship with the United Nations (Yugoslavia was a founding member). The publication ends with much shorter promotion text than that of the UK versions. In both English language editions, following the presentation of country's beauties and attractions, there is a testimonial by Nobel prize winner George Barnard Shaw:[1]

> WHAT G. B. S. SAID. – And if, in spite of all this, you still hesitate about coming to Yugoslavia, then listen to the authoritative voice of the greatest scoffer of England, to wit, the late George Bernard Shaw.
>
> Englishmen, Irishmen, Americans, and holiday-makers of all civilized nations, come in your millions to Yugoslavia. You will be treated like kings! The government will provide you with a perfect climate and the finest scenery of every kind for nothing. The people are everything you imagine yourselves to be and are not. They are hospitable, good-humoured and very good-looking. Every town is a picture and every girl a movie star. Come quickly before they find us out. It is too good to last. (Yugoslav National Office for Travel Promotion, 1953)

---

1   Note: in this text all citations are directly transcribed from brochures, including spelling and/or grammatical errors if any. All translations are made by authors.

The UK edition had another content that was not present in any other language editions. Under the heading A holiday in Yugoslavia, it is specifically said:

> For Yugoslavia is not only inexpensive, or, to be more precise, extremely cheap. (Yugoslav National Office for Travel Promotion, 1953)

The devaluation of Yugoslav dinar from 50 to 300 dinars for a US dollar in 1952, has made Yugoslav tourism very affordable and has initiated the continuous growth if foreign tourism. Low prices eventually become the most successful attractor to visitors of Yugoslavia and price competitivity remained county's tourism market strategy throughout its existence. The French and German brochures had also edition-specific text, but similar in content to the one in the UK edition (without the testimonial and the notion of low prices), with some obvious adaptation of text where it was country specific (visa procedures, access routes, exchange rates, etc.).

The production of brochures in separate language editions, mostly English, French, German and Italian, continued every year. Occasionally editions in other languages, such as Swedish of Russian have also appeared. In all subsequent publications there are hardly any variations in the written content, layout, or photographs among different language editions.

6 In search of uniqueness

During 1950s, the competition on tourism market in Europe started to grow. Yugoslav government understood the need to provide a competitive proposition to attract foreign visitors. In the attempt to distinct itself on a tourism market, Yugoslavia often promoted itself as "the land of contrasts". Tourism brochures were emphasizing the ethnic diversity of Yugoslav people, the beauties of the landscape and country's nature (from Alps to the Adriatic Sea) or being a country on a variety crossroads: historical and cultural (Habsburg and Ottoman), geographic (east and west, Alps and Mediterranean), ethnic (Slavic, Romanic and Germanic), political (capitalism and socialism). Folklore was also frequently used as differentiation element. Colourful national costumes and folk dances from all parts of the country were consistently represented in brochures and posters. Folklore evenings with traditional costumes and dances were part of standard entertainment program in tourist destinations across the country. The description of Yugoslavia as having "seven borders, six republics, five nationalities, four languages, three religions, two alphabets" (Pahor, 1965), all that in one country, was jet another attempt to distinct Yugoslavia from other countries by stressing its multi-ethnic population and reach cultural heritage.

The sunny Mediterranean climate was Yugoslavia's leading tourism product. Tourism demand was concentrated in two summer months (July and August) that contributed about half of the recorded annual tourism arrivals. Since 1954 the sea and seaside destinations were presented in a separate brochure The Sunny Adriatic (Sonnige Adria, L'Adriatico ridente di sole, L'Adriatique ensoleillée, Det soliga Adria, etc.).

Yugoslavia tried to attract foreign visitors not only with the sun and sea, excellent food, cultural richness, beautiful nature, or with centuries old historical monuments, but also with country's political system. In a 1957 brochure, foreign visitors were offered the opportunity to experience the creation of a new socialist society.

> Finally, should he wish it, the tourist will be able to get acquainted here with something that no other country in the world is able to show so clearly and unobtrusively – the creation and building up of a new social and economic system, which presents the practical consequence of a deeprooted humane idea. The same idea may be found reflected in the constant struggle of Yugoslavia for a peaceful and active coexistence among the peoples of the world, and therefore irrespective of its traditional hospitality, Yugoslavia will be glad to welcome visitors from all parts of the world, the point of view of the most diverse interests, because she is aware that by doing so she is con- tributing to international cooperation. (Čulić, 1957)

The text continues with a testimonial that reaffirms what has been said. The brochure cites Herbert George Wells (1866–1946), English science fiction writer and a Nobel prize nominee:

> This little publication is trying – modestly and briefly – to show the foreign tourist a few aspects of this ancient, though ever young country, with some of its natural and cultural characteristics.

> And if this invitation to visit Yugoslavia proves successful, the tourist will be able to confirm on the spot the words of H.G. Wells who said that "Yugoslavia does not only abound in beautiful scenery and historical monuments, but also in sincere cordiality and social atmosphere where there is no place for snobbery, pretention and affectation." (Čulić, 1957)

The brochures in general did promote the beauties of the country, but a real trigger for growth of the number of foreign visitors was the affordability of holidays spent in Yugoslavia (Stanković, 1990). Low price policy that ensured the competitivity of the country was openly promoted in tourist brochures. Here are two examples:

> If you want to spend your vacation inexpensively, visit Yugoslavia, of which you will keep unforgettable memories. (Federation touristique de Yougoslavie, 1959)

> Tasty food and a variety of delicious wines and other drinks as well as comfortable hotel accommodation make a stay on the coast very pleasant, although by no means expensive, even at the best hotels. (Tourist Association of Yugoslavia, 1963)

Often brochures also emphasized that the visit during off-season is even more affordable.

Hospitality of Yugoslav people was also exposed in tourism promotion as was county's constant struggle for peaceful cooperation among nations. At the end of the 1960s, already engaged in a fast growth of international tourism, the Tourist Association of Yugoslavia wrote:

> In such a remarkable country live a people proud of their past and the values created and maintained in the centuries-long struggle. Their folklore is full of colour, movement, music and joy of life, and their life today, as always, is filled with staunch aspirations for progress and peace.
>
> These people velcome guests, in accordance with ancient Slavic traditions, but they endeavour to express it in quite a modern way; by building roads, airports, harbours, ships, hotels and motels, by offering a varied cuisine and a large choice of drinks, by providing entertainment – from hunting, fishing and all kinds of water, winter and other sports, to folklore, music and drama festivals. If he needs it, the visitor can find relaxation in many spas and health resorts, and if he wishes, he can retire into the romantic peace of small tourist centres, far away from the bustle and noise of the present-day nomads. (Tourist Association of Yugoslavia, 1965)

7 A sea of brochures

Yugoslavia has understood the promotion potential of brochures in tourism promotion. To Yugoslav authorities conveying the image of an attractive destination for holidays was not less important than demonstration of professionalism in production of promotion publications. Engagement of marketing experts, artists, photographers, and text writers was common practice in design and production of promotion materials. Brochures, tourist maps, and posters were produced by best printers to assure final quality of publications that were destined to promote Yugoslavia abroad. Several books and manuals on tourism promotion were issued in an effort to assure the spread of knowledge to the regional, destination, or company levels, such as *Turistička propaganda* [Tourism promotion] (Čulić, 1965) or *Umjetnički izraz u turističkoj propagandi* [Artistic expression in tourist promotion] (Vukanović, 1968). Assessment of effectiveness of tourism promotion strategy was usually assessed through interviewing foreign tour operators and tourism officials as well as through regularly conducted visitor surveys. A research on effectivity of tourism promotion conducted among foreign visitors in Yugoslavia during the season of 1963 has clearly exposed travel brochures as most effective promotion tool. Brochures were recognized as being most trustful information source after, obviously, previous personal experience and friends' recommendations (Čulić, 1965). The results justified Tourist Association's focus of brochures as main tourism promotion tool.

During the 1950s and 1960s a standard yearly production of promotion publications by Tourist Association of Yugoslavia consisted of a 28-32-page booklet of tourism in Yugoslavia, a separate brochure promoting seaside tourism (The Sunny Adriatic), and a folded tourist map where the map was not a standard route of geographic map but rather a colourful drawing with illustrations of tourist attractions at specific locations. All publications were printed in several hundred thousand copies for each language version, and distributed through a network of representative offices, tourism organizations and tour operators. Besides that, the Association published other information booklets, brochures and posters for selected market segment or tourism products such as brochures promoting hunting tourism, camping guides, hotel directories, or brochures with suggested itineraries for self-driving visitors. Besides Tourist Association of Yugoslavia, Tourist Associations at the level of socialist republics also produced several publications promoting tourism within specific territory. Tourism destinations, travel agencies, hotels and other tourism companies have also contributed their share. During the 1950s, the production of promotion publications has already grown beyond control. A tentative compilation of a bibliography of publications issued by Tourist Association of Yugoslavia (or its predecessors) between 1945 and 1958 done in 1958 was not able to identify all published editions (Tchoukarine, 2010).

Almost all brochures and tourist maps edited by Tourist Association of Yugoslavia during 1950s and 1960s had a colourful artistically designed front page. The design frequently included symbolic representation of nature, urban centres, elements of cultural heritage or other attractions, rather than a recognizable destination or attraction. Such solution has alleviated the possibility of conflicts and disputes about selection or "unbalanced" representation of depicted destinations. Within the brochures, there was enough space to include more place-specific photographs. First brochures with a photographic cover appeared in late 1950s and at the beginning of 1960s. A 1959 8-page folder issued by Tourist Association of Yugoslavia entitled *Jugoslavija* had on the front page a photograph of a sailing boat on the lake Ohrid in Macedonia. Another booklet published in 1961 had a cover with a photograph of a beach scene where a young couple enjoys the sun and sea in an unidentified destination. A similar scenography was used for 1962 issue of *The Sunny Adriatic* brochure. However, photographic front pages were rare until 1970s, when photography prevailed.

By the end of the 1960s, the country was getting ready to become a more important player in European tourism. Economic benefits of tourism have become important to the national balance of payments, and tourism has received adequate attention and priority in allocation of investment funds. The fastest way to increase country's receptive capacities needed to be able to increase the volume of visitors, was the construction of camping sites. They needed less funds than the construction of a hotel and had produced immediate rise of available accommodation capacities. Even more effective solution was to stimulate citizens to rent rooms to tourists. This policy also produced immediate results in rising of available catering facilities, but without any investments from the state. The government has also started a series of hotel and resorts development projects, a program that has continued into the 1970s. During the 1960s the number of available hotel beds rose from 30,000 in 1960 to 120,000 in 1970, while the total number of available capacities has risen from 250,000 in 1960 to 700,000 in 1970.

The need for promotion publications was growing adequately. The needed volume of brochures had an impact of production costs that Tourist Association of Yugoslavia and other organizations responsible for tourism promotion were not able to sustain. International competition on tourism market has put pressure on costs and production-time of promotion publications that resulted in abandonment in established practices and required adaptation to new reality. The brochures were substituted by much cheaper folders. Graphic artists were excluded from the production process and their work was replaced by photographs, often of questionable quality. More often multi-language brochures replaced single-language versions. Printers were selected on a production-cost basis, that not always gave best results. However, as mentioned before, until prices were competitive, there was no fear of a drop in tourism demand (Stanković, 1990). In 1961 Yugoslavia recorded more than one million visitors from abroad. A decade later, in 1970, 4.8 million foreign visitors spent their holidays in Yugoslavia.

8 Conclusions

In about two decades of post-war tourism development, Yugoslavia evolved into a meaningful player in European tourism. At the beginning of this process, immediately after the war, tourism was not getting as much attention as other sectors of the economy (e. g. industry or agriculture) although it was not totally overlooked. The government has recognized social and cultural benefits of tourism and used it in the process of creation of a new Yugoslav socialist society. Distancing of Yugoslavia from Soviet Union led Eastern European block of countries forced Yugoslav government to consider opening of the country towards Western Europe to avoid the risk of isolation on the political map. Tourism was recognized as a viable option in this process. At first it was still used as political tool to convey "the truth" about new socialist system, but it was soon recognized as a source of foreign currency needed for the development of the country. State issued tourism promotion materials reflected changes in Yugoslav government's attitude towards tourism as well as changes in primary target markets for tourism. In immediate post-war period, tourism promotion was almost exclusively addressing domestic market. Promotion materials produced during this period were more educational brochures propagating the need for rest and holidays needed to regain strengths for the construction of a new society, than they were classic tourist brochures describing the beauty and attractiveness of the country. During 1950s, after Yugoslavia consolidated its rather unique political and economic system, tourism was recognized as a possible source of needed foreign currency. To increase the volume of visitors from Western Europe, Yugoslavia had to intensify its promotional activities on European markets. At first, the endeavour in international tourism was cautious, if not suspicious, and was not without political connotations. Tourism brochures for foreign visitors promoted county's nature and history but also its political and economic system. They mentioned fight for freedom and prized the achievements of new socialist society. Gradually, as country had assured its position on political map, the ideological content in tourism promotion diminished. Tourism promotion shifted from serving political agenda to serving commercial interests of the industry. During 1960s, in the process towards the massification of tourism, Yugoslavia positioned itself as a cheap country for a seaside summer holidays. Emerging tourism demand in post-war Europe responded immediately, and the number of international tourists started to grow. Tourism promotion has changed accordingly. Instead of describing country's richness and national treasures the promotion emphasized the benefits for a visitor. Brochures, cheaper and smaller in size, were limited to sell "sea, sand and sun." The following text from a late 1960s brochure sums it up:

> Finally if one does not care for any of those entertainments, one can relax – if one so wishes – and enjoy a »total rest« in whichever way one chooses. The selection is great: from the "dolce far niente" variety, by simply basking in the hot Adriatic sun on the sandy beaches of a small and peaceful fishing village, to the refined comfort offered by modern tourist trade which include night-clubs, several casinos, heated sea water swimming pools, and, of course, gastronomic delicacies. Let us add that an increasing number of hotels in many seaside places cater for a comfortable stay in winter months (the climate is so mild in winter that on a rainy day the guests in some of the hotels

pay only half the price of their' board and lodging and nothing at all, if by some freak of nature, it snows!). And a final word: the prices are favourable too. Outside the main season (which is in July and August) prices are reduced by as much as forty per cent. Let this prospectus, with its pictures and brief descriptions of the Adriatic resorts therefore, be like a small overture, and at the same time a cordial welcome, to THE SUNNY ADRIATIC. (Tourist Association of Yugoslavia, 1968)

References

Allcock, J. B. (1986). "Yugoslavia's tourist trade. Pot of gold or pig in a poke?." *Annals of Tourism Research* (13), 565–588.

Benić Penava, M. and Matušić, Đ. (2012). "Development of accommodation facilities in the Dubrovnik district between the two world wars: starting point for the development of modern tourism". *Acta Turistica* (24), 61-85.

Čulić, D. J. (1954). *The Sunny Adriatic*. Zagreb: Tourist Association of Croatia.

Čulić, D. J. (1957). *Jugoslavija*. Belgrade: Tourist Association of Yugoslavia.

Čulić, D. J. (1965). *Turistička propaganda*. Zagreb: Panorama.

Duda, I. (2010). "Workers into tourists. Entitlements, desires, and the realities of social tourism under Yugoslav socialism." In H. Grandits & K. Taylor (Eds.), *Yugoslavia's Sunny Side. A History of Tourism in Socialism (1950s–1980s)* (pp. 33–68). Budapest-New York: Central European University Press.

Duda, I. (2016). "When capitalism and socialism get along best: tourism, consumer culture and the idea of progress in Malo misto." In K. Taylor, I. Duda & P. Stubbs (Eds.), *Social Inequalities and Discontent in Yugoslav Socialism* (pp. 173–192). London and New York: Routledge.

Federation touristique de Yougoslavie. (1959). *Jugoslavija*. Beograd:author.

Kobašić, A. (1987). *Turizam u Jugoslaviji*. Zagreb: Informator.

Kranjčević, J. (2019). "Tourism on the Croatian Adriatic coast and World War I." *Academica Turistica* (12), 39–50.

MacCannell, D. (1992). *In Empty Meeting Grounds: The Tourist Papers*. London: Routledge.

Pahor, B. (1965, Nov. 29). "Britanci odkrivajo Jugoslavijo." *Tovariš*, 21(47–48), 44–46.

Pirjevec, B. (1988). *Ekonomski aspekti jugoslavenskog turizma*. Zagreb: Školska knjiga.

Putnik. (1949). *Trudbenici koristite odmor*. Beograd: Author.

Radišić, F. (1981). *Turizam i turistička politika*. Pula: Istarska naklada.

Repe, B. (1996). "Turizma ni mogoče zavreti, čeprav bi ga prepovedali z zakonom." In F. Rozman & Ž. Lazarevič (Eds.), *Razvoj turizma v Sloveniji: zbornik referatov* (pp. 157–164). Ljubljana: Zveza zgodovinskih društev Slovenije.

Sitar, P. (2020). "Workers becoming tourists and consumers: social history of tourism in socialist Slovenia and Yugoslavia." *Journal of Tourism History*, (12-3), 254–274.

Službeni list. (1946). "Uredba o plaćenom godišnjem odmoru radnika, namještenika i službenika." *Službeni list*, 2(56).

Službeni list. (1973). "Zakon o međusobnim odnosima radnika u udruženom radu." *Službeni list*, 22.

Stanković, S. M. (1990). *Turizam u Jugoslaviji*. Beograd: Turistička štampa.

Taylor, K. and Grandits, H. (2010). "Tourism and the making of socialist Yugoslavia". In H. Grandits & K. Taylor (Eds.), *Yugoslavia's Sunny Side. A History of Tourism in Socialism (1950s–1980s)* (pp. 1–30). Budapest and New York: Central European University Press.

Tchoukarine, I. (2010). "The Yugoslav road to international tourism. Opening, Decentralization, and propaganda in the early 1950s." In H. Grandits & K. Taylor (Eds.), *Yugoslavia's Sunny Side. A History of Tourism in Socialism (1950s–1980s)* (pp. 107–138). Budapest and New York: Central European University Press.

Tchoukarine, I. (2015). "Yugoslavia's open-door policy and global tourism in the 1950s and 1960s." *East European Politics and Societies and Culture* (29), 168–188.

Tmušić, D. (2013). Reprezentacija Zapada u jugoslovenskoj turističkoj propagandi 1970-ih. Beograd: Forschungsprojekt „Repräsentationen des sozialistischen Jugoslawien im Umbruch" (Working Papers, 13).

Tourist Association of Yugoslavia. (1960). *Yugoslavia*. Belgrade: author.

Tourist Association of Yugoslavia. (1963). *The Sunny Adriatic*. Belgrade: author.

Tourist Association of Yugoslavia. (1965). *Yugoslavia*. Belgrade: author.

Tourist Association of Yugoslavia. (1968). *The Sunny Adriatic*. Belgrade: author.

Vukanović, Đ. (1968). *Umjetnički izraz u turističkoj propagandi*. Beograd; Turistička štampa.

Yeomans, R. (2010). "From comrades to consumers. Holiday, leisure time, and ideology in Communist Yugoslavia." In H. Grandits & K. Taylor (Eds.), *Yugoslavia's Sunny Side. A History of Tourism in Socialism (1950s–1980s)* (pp. 69–106). Budapest-New York: Central European University Press.

Yugoslav National Office for Travel Promotion. (1953). *Yugoslavia Awaits You!*. Belgrade: author.